Management of Engineering Projects

Victor G. Hajek
Project Management Consultant

THIRD EDITION

McGraw-Hill Book Company
New York   St. Louis   San Francisco   Auckland
Bogotá   Hamburg   Johannesburg   London   Madrid
Mexico   Montreal   New Delhi   Panama   Paris
São Paulo   Singapore   Sydney   Tokyo   Toronto

Library of Congress Cataloging in Publication Data

Hajek, Victor G.
    Management of engineering projects.

    Includes index.
    1. Engineering—Management.   2. Industrial project
management.   I. Title.
TA190.H35 1984   658.4'04   83-17514
ISBN 0-07-025536-9

1 2 3 4 5 6 7 8 9 0   DOC/DOC   8 9 8 7 6 5 4 3

TA
190
.H 35
1984

ISBN 0-07-025536-9

The editors for this book were Diane Heiberg and Olive H. Collen,
the designer was Naomi Auerbach, and the production
supervisor was Teresa F. Leaden. It was set in Caledonia
by University Graphics, Inc.
Printed and bound by R. R. Donnelley & Sons Company

CONTENTS

PREFACE

During the past decade, rapid technological advances have made possible the development of sophisticated systems which are capable of meeting previously impossible military and industrial performance requirements. The most dramatic achievements involve the application of the new generations of computer equipment.

The implementation of the new technology, the management of the activities related to the new designs, and the efficient utilization of scarce, highly trained engineers, computer scientists, programmers, and other professionals have presented new challenges to project managers. In addition, changes have taken place in areas such as procurement procedures and monitoring and control techniques with which project managers must be familiar in order to perform their functions. To address the new and revised responsibilities of project management, the text of *Management of Engineering Projects* has been revised, and additional subject matter included, to reflect those new responsibilities and the changed roles of individuals responsible for today's projects.

One of the many new areas covered in the text relates to the functions the project manager must implement to effectively manage the development of computer software. The latest management procedures for monitoring and controlling the computer programming effort are also presented. The coverage and discussion of these and other subjects are achieved through the use of a case study of a project involving a computerized simulator wherein the different responsibilities of the project manager—from the time of proposal preparation, through the various project contract phases, and to the delivery of the finished product—are illustrated.

In attempts to realize the efficient utilization of resources, promote a companywide team effort, provide for high visibility of all projects by top management, and achieve a more flexible type of operation, many companies have adopted different forms of the matrix type of organization. The principles of the matrix and the differences between the matrix and the traditional line organization are presented. Since the functions, responsibilities, and authority of the project manager in the matrix are significantly different from those in the line organization, the text discusses the revised role and attitudes the project manager must adopt when functioning in the matrix.

The book includes figures and tables that supplement the narrative; generally the details of illustrations were edited so that the management principles being discussed could be clearly presented. In addition, bibliographies are included at the end of most chapters to permit the reader to pursue in greater depth the subject matter that is presented.

The effort to complete the book required assistance from many sources. In particular, I would like to acknowledge and express appreciation to D. Wilson, R. Wright, C. Blake, G. Carter, and R. Jarvis, who provided valuable input during the creation of the text.

It is hoped that the material in this book will serve as a useful guide for engineers, managers, and individuals who have an interest in the important profession of project management.

Victor G. Hajek

THE MANAGER IN PROJECT ORGANIZATIONS

1.1 Line and Matrix Organizations

Recent technological advances have made possible the creation of complex systems that previously were not feasible. The development of such systems involves the utilization of expensive manpower, equipment, facilities, and other resources that are in high demand and are often not efficiently utilized in traditional line-project organizations. In many cases the use of the matrix system has resulted in the reduction or elimination of the resource problems that are inherent in line organizations.

In addition to the improved utilization of resources, other advantages, relating to more open communication and greater project visibility to management, are provided by the matrix. Since many project managers are involved with the type of system development to which a matrix type of organization would apply, they must be prepared to function in a management environment that might be fundamentally new and foreign to them.

Project managers who have achieved a track record of success while supervising projects in a line organization often are not able to adapt when charged with performing the same type of management functions in a matrix organization. The reasons for this apparent paradox are related to some degree to the differences in the types of authority and modes of operation that the two organizations demand. The difficulties that many managers experience with the matrix are due to personality traits and to attitudes that they possess which are incongruous with what the matrix organization requires. Specifically, the matrix project manager must consider how decisions contemplated for an assigned project will affect the total organization rather than how they will affect the isolated world of one project. The ability to view an assigned project and the parent organization in proper perspective and to make or accept decisions accordingly requires such special personal characteristics as flexibility, objectivity, and communication ability.

The discussion as to when it would be more advantageous for a company to function under a matrix as opposed to a line structure is a broad subject of its own. It should be sufficient to note that the transition to the matrix structure is common and that the project managers of today must be prepared to function in either type of organization. Because of the prevalence of the two types of organization, both the line and the matrix will be addressed for those project manager functions which are unique to each type of structure.

The traditional line (or functional) structure is based on a pyramid or hierarchy command common to military, church, and similar organizations that have existed throughout history. The common structure is depicted in Figure 1.1. The hierarchy of authority descends from the single highest point and spreads out through the lower echelons. In Figure 1.1, the manager designated as B has unilateral authority over individuals D, E, and F in the branch. However, manager B has no authority over individuals G, H, or I and, strictly speaking, communicates with the other managers in the same echelon (C) only through A.

Thus the line organization tends to breed a provincial environment for the project managers and members of each of the project teams which often does not serve the best interests of the parent organization.

A project organized along the line structure would have an appearance similar to the simplified example shown in Figure 1.2. In establishing the project team, all the resources deemed necessary for the project are assigned to the

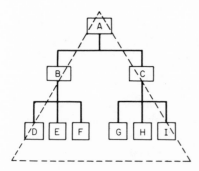

FIG. 1.1 Traditional line organization.

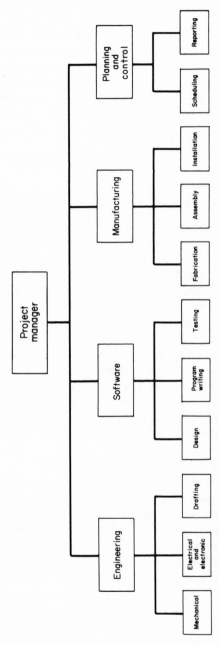

FIG. 1.2 Project structure in line organization.

team, and the project manager is responsible for meeting the project objectives with these resources.

The line project team is generally successful in carrying out its mission and meeting the schedule, cost, and performance objectives if no major difficulties are encountered and if project revisions are not imposed or required. However, many projects require the creation of a product designed to meet new requirements, and hence the development of new systems. Examples of circumstances that may involve substantial change are:

1. Changes directed by the customer
2. Revision of design approaches by management because of engineering problems
3. Management redirection of design approaches to take advantage of new technology that is available

Customer-directed changes involve amendments to any or all of the contract objectives of cost, delivery, and equipment performance. When such changes occur, the project manager must accurately appraise the impact of the changes and restructure the team resources to make sure that the types of personnel required to implement the revised objectives are included and that the project cost and schedules are realistic.

Internally directed changes usually result from misjudgments by the management echelons of the scope or complexity of the project or from errors which require the pursuit of new approaches. It is the project manager's responsibility to direct corrective actions with the resources of the project team.

The adjustments in workload schedules that are necessary to implement changes often require peak efforts for the existing personnel resources of the project team, especially if the original equipment delivery dates are not revised. Project managers find that the production efficiency of engineers and other personnel who work 70- and 80-hour weeks deteriorates rapidly in a short time and that the entire project plan begins to fall apart.

If the existing resources of the project team are not sufficient to implement the necessary revisions, the line project manager can seek authorization to have personnel from other projects reassigned, to hire job shop personnel, or to enter into subcontract arrangements in order to implement corrective actions. However, any particular course of action will exact penalties of cost, time, rescheduling of other projects, or some other undesirable result that the parent organization must assume. In addition, such remedial actions often require that the project objectives be renegotiated with the customer. It is the responsibility of the project manager to prepare the case for the renegotiation to avoid or minimize any adverse results that might be detrimental to the parent company.

Because technological advances prior to the 1970s occurred at a more leisurely rate, and because design approaches that were established were not subject to the scope and frequency of revisions that often characterize contemporary developments, the traditional line structures of project teams were successful and continue to be effective for most projects not involved in development effort or subject to revisions. However, modern complex systems that are being developed with the latest technology are generally more dynamic as far as revisions are concerned and require a flexibility of management that cannot be realized with a line organization. In addition to the difficulties that the project manager often experiences due to its inflexibility, the line structure tends to breed isolation among the different projects in the parent organization. Project teams often compete with each other to the point where communications among project managers are throttled and project personnel have lost sight of the fact that the interests of the parent organization should be of primary concern rather than the interests of the project team. The matrix system evolved to remedy this type of problem in the project organization.

The structure and function of the matrix organization tend to eliminate traits that can handicap the line organization. In the case of a matrix project organization, the project manager is still charged with the responsibility of the project's success but does not have the autonomous authority that is inherent in the line organization. The acceptance of this anomaly and adjustment to the new mode of operation can be major impediments to the ability of a project manager to transfer from a line to a matrix structure.

The basic structure of a matrix project organization is illustrated in Figure 1.3.

The two major differences between the line structure and the matrix structure, as far as the project manager is concerned, are that in the matrix the pro-

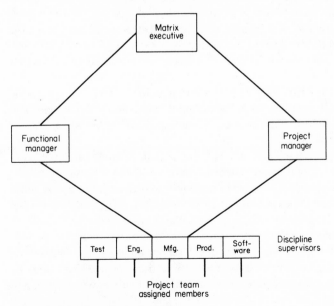

FIG. 1.3   Matrix project structure.

ject manager is not at the apex of the organization and that authority over the project team members and the discipline supervisors is shared with the functional manager. However, as previously noted, the project manager retains the responsibility for meeting the performance, cost, and schedule objectives of the project. The cooperative execution of the responsibilities of the members of the matrix team is designed to provide a team operation which serves to remedy the major line project team shortcomings of inflexibility, team isolation, communication difficulties, and provincial attitudes towards individual projects.

1.2 Managers in the Matrix

The parties of the matrix with whom the project manager must interface and work out solutions to problems are depicted in Figure 1.3.

In order to function effectively in the matrix, the project manager must understand the role and responsibilities of the other matrix members. The role and responsibilities of the key managers of the matrix can be summarized as follows.

1. *Matrix executive:*   The head of a project matrix organization is identified by any of a variety of titles, including chief executive, president, and matrix executive. The term "matrix executive" appears to best define the individual whose major responsibilities include the following:

 a. Making sure that neither of the two legs of the matrix triangle noted in Figure 1.3 dominates the structure. If either the functional manager or the project manager were able to influence the other, the basic feature that enables both parties to negotiate as equals and to objectively achieve a compromise that best serves the interests of the project as well as of the parent organization would be destroyed. It is the responsibility of the matrix executive to make sure that a balance of power is maintained in the matrix structure.

 b. Making sure that open communication is maintained among matrix managers. Candidness and open discussion of problems, achievements, and other matters are basic characteristics of the matrix and must be maintained by the matrix executive.

 c. Functioning as the moderator of differences in an environment of "controlled conflict." When issues cannot be resolved at the manager level, the matrix executive assumes the role of arbitrator and provides whatever additional relevant information is required to achieve a compromise course of action that parties to the conflict can accept and implement.

 d. Establishing objectives and levels of resource commitments for areas of the project that require research or development effort. Those engineering objectives and the resources required to achieve the objective that cannot be quantified from previous experiences represent an unknown. One of the responsibilities of the matrix executive is to make a judgment as to what objectives are to be achieved and to establish the amount of resources to be expended.

2. *Functional manager:*   The functional manager is the counterpart of the project manager and heads one of the legs of the matrix structure shown in Figure 1.3. The primary responsibility of this manager is to control and provide the resources that are required for the various projects in the parent organization, including those required by any specific project manager. Some of the major duties that the functional manager is required to carry out are as follows.

a. Negotiate with project managers regarding resource requests.

b. Monitor and control utilization of resources by projects in the parent organization.

c. Maintain even distribution of resources.

d. Review, negotiate, and carry out revisions to resource requirements of projects as necessary.

e. Participate with project managers in evaluation, assignments, and other personnel matters relating to discipline supervisors and their assigned team members.

f. Handle union negotiations and other related functions.

3. *Discipline supervisor:*   The discipline supervisors represent the third echelon of the matrix structure. These supervisors and their team members constitute the "two boss" individuals, since they ultimately are responsible to both the functional and project managers.

As a group leader, each discipline supervisor must be able to function in an organization requiring allegiance to two bosses. Special personal qualities of flexibility, articulateness, and communication abilities are required. The discipline supervisor's responsibilities include the ability to effectively perform the following functions.

a. Direct and carry out the responsibilities of the team.

b. Communicate with both managers to promote desired courses of action.

c. Analyze the positions of each manager on issues so as to be able to promote desired approaches with the most effectiveness.

d. Establish "fallback" positions that represent acceptable compromises.

e. Support each of the members of the discipline groups when discussions about performance, discipline. or other personnel matters are conducted.

Figure 1.4 is an expansion of Figure 1.3 and depicts the "two boss" relationship of the members of the discipline groups when two or more projects are pursued in an organization. The members of the discipline groups in any area can be assigned to more than one project. For instance, an engineer might be required to perform tasks in both discipline team A and discipline team B for projects handled by project managers I and II. The scheduling of the efforts of

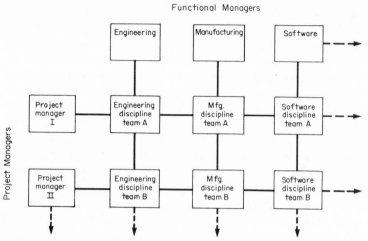

FIG. 1.4   Multiple projects in matrix organization.

particular engineers would have to be resolved by the managers of the assigned projects and the functional manager responsible for the engineering resources.

1.3   Project Manager Functions

The project manager is responsible for managing the technical project in a manner that will result in meeting the project objectives. Most projects involve a contract between two or more parties. In addition, an organization may charge the project manager with the responsibility of an "in-house" technical effort designed to achieve a particular end result, such as developing a new project or system that can be marketed. In this text, the type of project involving a contracted effort and the role that the project manager fills in managing a contracted project will be discussed. To illustrate the many functions of the managers and the wide scope of disciplines that must be applied, a case study will be presented involving government procurement of a radar landmass simulator (RLM simulator) contracted from a fictitious company, Creative Electronics Corporation.

The RLM simulator was chosen because each new generation of the device involved engineering development effort and utilized the latest technological design available at the time. Since some of the greatest challenges to project managers relate to creative engineering effort and the effective application of new technology, the use of the RLM device is very appropriate for illustrating various principles and procedures related to the subject of managing projects.

Devices are required to train operators in the use of radar equipment and to develop the ability to interpret landmass presentations on the radar display. The earliest designs were relatively crude electromechanical devices which were adequate for the type of radar equipment available at the time. With the creation of more sophisticated radar systems, new simulators were developed which utilized the latest applicable technologies available. Because of the advances made in digital computer technology and programming techniques, computer systems represent the new generation of RLM simulator designs, superseding the electro-optical systems using transparencies. The text will address the project manager's role in the analysis of feasible technologies, leading to the choice of the computer approach and the managing of a contract for a digital computer RLM simulator. Engineering details will be presented only to the extent necessary to illustrate the technical decision-making responsibilities of the project manager. Because the project manager's responsibilities embrace not only technical but also such non-technical areas as contract administration, legal issues, and other "business" disciplines which are vitally important to the success of the program, much of the text will also cover such areas as scheduling and contract law. Differences in the project manager's responsibilities when functioning in line as opposed to matrix organizations will also be presented.

The three basic objectives for which the contractor's project manager is responsible are as follows.

1. Delivering a product that meets the requirements of the specification
2. Delivering a product that meets the requirements of the contract delivery schedule
3. Meeting the company's profit objectives for the contract

The following are the prerequisites that must be satisfied if the foregoing objectives can be expected to be achieved.

1. The particular engineering, production, and management resources required for meeting the equipment performance objectives must be available to the company. For example, a company would be hard-pressed to design and produce equipment incorporating a digital computer if the company did not possess or have access to experienced programming personnel.
2. The company must have available adequate resources and facilities to meet the delivery schedule of the contract.
3. The terms of the contract (price, rates, etc.) should be realistic and adequate to cover what it will cost the company to live up to the terms of the contract.

Project managers have the dominant role in identifying the resources required to meet the objectives, since they are involved with the project from the time the request for proposals is analyzed through the equipment acceptance and often through the contract close-out effort. The achievement of these objectives requires that the project managers be proficient in a scope of disciplines covering managerial as well as engineering functions. The major disciplines and the functions which would be applied for the line and matrix organizations are summarized below.

1. *Engineering:*   By its nature, the project-oriented task cuts across several areas of engineering effort. A military weapon system might require expertise in such disciplines as electronic, electrical, and mechanical engineering. If the project was for a processing plant the required disciplines would include chemical, electrical, electronic, and mechanical engineering. In addition, new categories of engineering and scientists have evolved during recent years. The one major group which performs a vital function in contemporary technology consists of computer scientists and programmers. Project managers cannot be expected to be expert in all the technical areas that a project requires, but they must be familiar enough with the basics of different technologies to consider alternatives that the engineering specialists may offer and be able to evaluate and make correct decisions in the various areas.

2. *Cost management:*   All the various areas for which project managers are responsible can be ultimately reduced in most cases to one common denominator: money. If a particular engineering design approach proves to be in error, the necessary redesign will involve unanticipated expenditures of money. If the time to produce a product exceeds the planned schedule, additional cost will be involved, since the time that equipment remains in a plant beyond its scheduled shipping date represents unanticipated cost expenditures to the contractor.

The discipline of cost management, which is one of the prime responsibilities of project managers, embraces the following functions:

Cost estimating
Project cash flow
Direct and overhead rate controls
Incentive, penalty, and cost-or-profit-sharing considerations

3. *Contract law:*   A project involving a buyer and a seller is almost always based on a legal agreement or contract between the participating parties. Project managers, who are assigned the responsibility to have the employer's contractual obligation fulfilled, must be fully knowledgeable of the contract terms as well as sensitive to the implications of these terms. Normally both the procuring activity and the contractor have project managers and chiefs charged with the responsibility of fulfilling their employers' contractual obligations. As noted earlier, this text will deal primarily with the project manager representing the contractor.

4. *Negotiations:*   Most larger organizations employ specialists trained in negotiation psychology, tactics, procedures, and standards of conduct. However, the specialists rely on the project manager to assist in negotiation conferences and to provide detailed technical information. As a participating member of the company's negotiation team, the project manager must be aware of what and what not to say, when silence should be maintained, and when to speak up.

In smaller organizations, project managers often are the company's primary individuals in negotiations. In either case, they represent a key factor in determining the success of their company's efforts in obtaining favorable contract awards.

5. *Scheduling:* A contract for the design, fabrication, and delivery of equipment for which the project manager is responsible almost always involves complex scheduling problems. The proper phasing and scheduling of the different types of effort required by the project are essential to the meeting of the contract delivery dates. Project managers must therefore be familiar with the various engineering tasks, processes, and available resources which are necessary for executing the contract.

In the line organization, the project manager establishes and monitors the schedule of the project personnel and resources as judged appropriate for meeting the objectives. The matrix project manager does not possess the unilateral control over the scheduling of the resources required to execute the project, since he or she is assigned to the project as a result of schedules previously arranged with the functional manager. Thus if circumstances necessitate significant deviations from the existing workload schedules, revisions must be worked out during discussion and negotiations between the two matrix managers.

In order to effectively monitor and control the project schedules, both the line and the matrix project managers must be able to apply the various tools and techniques that are available to implement effective scheduling of the project. Some of the management tools that might be used include:

Program evaluation review technique (PERT)
Line of balance (LOB)
Gantt charts

1.4 Authority of Project Managers

A company or organization will make the decision to pursue a particular program when the management officials decide that the benefits that will be realized by achieving the objectives warrant the risks and effort involved. Some of the benefits that attract a company or organization are profit, the development of a new product that could result in profit, and the creation of a usable end item to serve a need.

When the decision to pursue a program as a project-oriented effort is made by management, the officials will select a project manager, who will be charged with the responsibility for achieving the program objectives and who will be provided with the authority to carry out these responsibilities.

The authority given to a project manager is broad in scope and embraces the disciplines discussed in Section 1.3. Some of the more common areas of decision-making authority of the line and matrix managers are as follows.

1. Technical decisions (the authority is essentially the same for both line and matrix managers)
   a. Directing design and computer programming approaches
   b. Selecting subsystems or components
   c. Identifying types and the scope of tests
   d. Selecting programming tools to be used
2. Commercial decisions (in the matrix, coordinated with the functional manager)
   a. Make-or-buy decisions
   b. Identifying subcontractors and vendors
3. Administrative decisions (in the matrix, the original plan and any subsequent revisions to resource requirements are established by agreements with functional managers)
   a. Selecting and assigning personnel
   b. Scheduling resources
4. Monetary decisions (the authority is the same for both line and matrix managers once a budget is established for the project); expenditures of project funds

1.5   Responsibilities of Project Managers

In the project manager's efforts to achieve the objectives of the project or program, the one thing that top management cannot tolerate is an unpleasant surprise such as an unreported technical problem or an unforeseen expense. Therefore it is essential that project managers report accurately, factually, and promptly all problems, errors, and potential problems.

Because of the inherent characteristics of the line project team, communication between the team and external parties tends to be inhibited. Status reports that are rendered periodically may prevent or minimize developing problems. In efforts to protect a professional or project image, managers sometimes try to work out problems with the available resources rather than follow the prudent path by seeking assistance from sources external to the project organization. One of the advantages of the matrix is that projects possess high visibility, and problems thus quickly surface.

Those line organizations project managers who are able to view the assigned project as a subset of the total parent organization and who can candidly communicate status information to serve the best interests not only of the project, but of the total corporate organization, are ideally suited for matrix project management assignments. The very nature of the matrix operation mandates that project managers be able to freely communicate and, in the overall interests of the parent organization, have the flexibility to adjust their project plans as dictated by higher authority.

In the case of a contractor's project objectives, project managers in both line and matrix organizations must maintain a close control and scrutiny of the funds that are involved in the program. Some of the monetary responsibilities of project managers are as follows.

1. Making judicious expenditures
2. Maintaining adequate cash flow into the program by meeting progress-payment milestones
3. Avoiding circumstances that might necessitate special funding to keep the program moving
4. Keeping profit objectives visible
5. Avoiding errors

The carrying out of the project managers' responsibilities requires that they monitor and report the technical, delivery, and financial status of the program frequently and in a timely manner. The most critical requirement of any report to management is that it must "tell it like it is." Any attempt to gloss over actual or potential difficulties is bound to be exposed at a later date, almost always too late for effective remedial action.

1.6 Summary

The project manager in a line organization has essentially autonomous authority as to the utilization of the resources that are assigned to the project for meeting the project objectives. In the matrix organization, the project manager is allocated resources for the project by a functional manager with whom authority over the project resources is shared.

The line project manager is expected to work out problems with assigned resources, whereas problems in the matrix are immediately surfaced and the project manager must communicate the problems to other matrix managers and negotiate resources for remedial action. Thus the matrix is better suited for projects which involve development effort and which are susceptible to changes of requirements or redirection.

In most cases, the program that the project manager monitors includes a contract between the procuring organization and the contractor. Project managers are assigned the responsibility for carrying out the customer's objectives and the contractor's objectives. Regardless of whether the line or matrix project manager represents the customer or the contractor, he or she must have a grasp of the fundamentals that cut across a broad spectrum of disciplines, including the various branches of engineering cost management, contract law, procurement regulations, negotiation techniques, scheduling procedures, management presentation, and other areas of administration.

BIBLIOGRAPHY

Davis, Stanley M., and Paul Lawrence: *Matrix*, Addison-Wesley, Reading, Mass., 1977, pp. 1–24, 37–52, 69–91.

2

THE PROCUREMENT PROJECT INITIATED

2.1 Solicitation

The Creative Electronics Corporation first becomes officially involved as a potential offeror in a procurement when the company receives the solicitation for the development of an RLM simulator. Since development projects usually utilize negotiated procurement approaches that require the submission of technical proposals, the text discussion will be based on the use of an RFP (request for proposal).

By definition, a procurement based on a performance specification allows consideration of a system based on any type of technology if the system can meet the requirements that have been identified. However, for reasons of logistics, commonality, future growth, etc., the using activity will often identify the type of technologies that will be considered. For instance, a specification calling for an electronic system would eliminate any option of an offer of a electromechanical system. Occasionally, options to offer one of two or more acceptable approaches are permitted in the specification. For example, as an illustration of the preliminary analysis procedures that the project manager must follow in establishing the design approach that will be offered, the procurement for the RLM simulator permits either a system based on computer technology or the more conventional electro-optical system that has been used in the past.

The RFP contains the information necessary for the offeror to prepare the technical proposal and the price quotation. The primary documents contained by an RFP include the following.

1. *Covering letter*:   Forwards the document relating to the procurement and gives a general description of the items to be purchased, the type of contract solicited, and any other basic information of interest.

2. *Statement of work (SOW)*:   Defines efforts required by the contractor and the requirements that are not otherwise covered in the specification. The SOW also provides information intended to enable offerors to make the decision as to whether to submit a proposal; sometimes it serves as a specification.

3. *Specification*:   Stipulates the performance and physical requirements for the primary equipment and other articles under procurement.

4. *Contract schedule*:   Indicates line items to be provided, delivery and other milestone dates, contract clauses, incentives and/or penalties, and any other contractual terms that are applicable to the proposed procurement.

5. *Technical proposal requirements (TPR)*:   Identifies the specific information that must be provided by offerors in their proposals. The details of the TPR are derived from the specification and serve to guide offerors as to the desired content of their proposals, thereby avoiding proposal material which the contracting party neither desires nor intends to evaluate.

6. *Cost breakout* (for negotiated procurements):   Indentifies the depth of cost detail that is required in the cost portion of the proposal. The detail can range from a single number for each line item that is listed in the schedule to a breakdown of cost for each kind of discipline (e.g., mechanical engineering, testing, and sheet metal processing). The cost breakout would normally include labor rates, overhead, G&A (general and administrative) rates, and profit.

7. *General and administrative clauses*:   Information relating to the procuring organization's policies. The offeror must indicate specific concurrence with the clauses in the proposal submitted. Typical of such information are patent rights, prohibition of prison labor, and employment of minority personnel.

8. *Data*:   Supplementary information to aid the offeror in identifying the articles to be procured. Data could be in many forms, including drawings, sketches, and publications.

2.2 Participation Decision

The preparation of technical and cost proposals, attendance at preproposal meetings, involvement in negotiations, etc. require the commitment of significant resources and money by a company. Since only one company will be awarded the contract, the funds and effort normally represent a loss to companies that have not been successful, and the proposal costs are charged to the company's overhead. Therefore, the decision to participate or not participate in a certain procurement must be based on careful analysis of the circumstances surrounding it and an evaluation of the risks that the company must be willing to accept in the interests of being successful in receiving a contract award.

The schedule of deliverable line items provides information as to what is required and the quantities of each item to be furnished. In addition, various sections of the contract schedule provide detailed descriptions of the line items, delivery schedules, and all the other terms and conditions of the procurement.

The SOW is a relatively recent document that is provided to offerors in many acquisitions. The SOW in many instances supplements the specification, schedule, and other acquisition documents, but it also adds an important new dimension to the information previously furnished which can be used by offerors in making the decision as to whether to participate in the procurement. After the award, the SOW serves as a list of performance criteria to be used by both the contractor and the procuring activity.

Some of the major types of information included in the SOW are as follows:

1. Contract scope and objectives
2. Contract end items
3. Requirements for data
4. References to other documents, studies, etc. that are relevant to the acquisition
5. Equipment, documents, data, etc. to be furnished by the procuring activity
6. Milestone schedules

Even with the assistance of the SOW information which definitizes the requirements of a procurement, qualified companies often decide against participation in a procurement due to the risk involved in committing large sums for the proposal effort.

Because of the critical importance of certain major defense procurements, the government has implemented a procedure for identifying the most qualified companies in a particular field and for entering into contracts with these companies, which then respond with proposals for a system under consideration. Companies who receive such contracts are therefore reimbursed for their proposal effort. The choice of different options enables the procuring agency to make a contract award for the offering that best serves the government's interests.

The individual who will function as the project manager on a program is usually not officially designated until the decision to participate is made by management. However, the project manager-designate will participate in pre-proposal discussions and provide inputs as appropriate and will continue in that capacity after the company desires to pursue the project. The points that must be considered in deciding whether or not to participate in a procurement were alluded to in Section 1.3 and are summarized as follows.

1. *Technical resources:*   Does the company possess or have access to the type and depth of engineering technology that the design of the items under procurement requires?

2. *Facilities:*   Are the available space, equipment, testing and other facilities adequate for the program?

3. *Workload:*   In the event of a contract award, would the company be able to handle the workload demands in addition to the existing workload?

4. *Competition:*   Does some other company have a significant advantage due to prior experience, expertise, patents, proprietary information, or political influence?

5. *Delivery:*   Are the delivery requirements reasonable and within the company's ability to perform?

6. *Risks:*   Does the procurement pose any significant risks to the company in such areas as technical difficulty, penalties on performance or delivery, and type of contract?

If after considerations of these points the company decides to budget funds, prepare a proposal, and participate in the preaward effort, the proposal team will be organized.

It is during the proposal preparation stage that the first indications of the differences between the functions of line and matrix project managers become apparent. In the line organization, the project manager makes the major decisions relative to proposed technical approaches, types and quantities of resources to be committed, scheduling, and other factors that are relevant to the procurement.

In the matrix, the decisions relative to the procurement elements are derived primarily from a joint effort of the functional and project managers. A technical approach established by the project manager may require modification due to necessary resource constraints that are established by the functional manager. Similar compromises would characterize the various subsystems of the equipment under procurement. Thus the autonomous authority of the project manager in the line organization is in sharp contrast to the team function that the project manager has in the matrix organization.

2.3   Technical Options

The technical role of project managers begins in earnest when they start to analyze the procurement specification, the TPR, and other documents. They must be intimately cognizant of the specification details to understand what must be designed, developed, and fabricated; and they must know the details cited in the proposal requirements document in order to effectively address those points which will be evaluated by the procuring officials. It is at this point that they must rid themselves of any preconceived ideas as to what is required and carefully study the specifications and proposal requirements with objective and open minds. All too often large expenditures of effort and money are made in preparing a proposal which reflects what the project manager would like or what he or she is best qualified to offer rather than what is required in the specifications. Unless the proposal addresses the points cited in the TPR document and provides information that will be judged superior to what the competition offers, the contract could be lost to a competitor.

A specification to be used for a procurement may be written as a design document which defines the required equipment in detail, or it may be written as a performance document which merely describes the end use of the product without defining the method, means, or design concepts to be used. Thus a design specification provides engineering information so that the ultimate contractor is not required to create an original design. The contractor has to possess the ability to interpret and translate the information in the drawings and design documents into the hardware. A performance specification, however, requires creative engineering effort. The more general the performance specification is, the greater the imagination that must be applied for the successful execution of the program. Since the role of the project manager usually involves the administration of contracts based on performance specifications, this text will confine itself to analyzing and resolving the problems commonly encountered in procurements based on such specifications.

2.4 Specification Terms

During the course of a year, many different kinds of performance specifications are written for equipment of many types, and thousands of companies submit proposals with hopes of obtaining contract awards. Except for the specific technical knowledge required, the role and functions of the project manager are essentially the same for any project. Although the case discussed in this text relates to a computer system that is offered on a government procurement, the principles discussed and the logic with which the decisions are made apply to practically any field of endeavor in which a project manager may function. It should be noted that the information is also applicable to procurements by commercial customers and corporations.

Upon completion of the review of the specification, the project manager should prepare a summary of the major and significant details of the project requirements. The following type of summary will serve to identify and keep visible the major requirements of the project for both the project manager and the other management officials in line or matrix organizations.

SUMMARY OF MAJOR REQUIREMENTS
SPECIFICATION No. 1001
DATED 1/15/83

Radar Landmass Simulator—Device 5A1

1. Scope.

   a. Radar Landmass Simulator—Device 5A1 is to be designed, developed, and constructed to function with aircraft flight simulator A4K, serial 9275.

   b. Simulated radar return from any preselected geographical area is required.

   c. The position and movement of the aircraft to be simulated are needed.

   d. The design approach is either digital-computer or electro-optical. Each of the offerors is to propose one or the other.

   e. Requirements for performance accuracies, reliability, size, and testing are specified.

   f. The system design approach is limited to electro-optical design using transparencies as data or to digital computer system design with programmed data.

   g. For digital computer designs, programming language shall be FORTRAN IV.

2. RLM simulator requirements.

   a. AN/APO-28 airborne radar is to be simulated and should function with the existing flight simulator.

   b. Simulated parameters.

   (1) Altitude   0–20,000 feet

   (2) Speed   250–600 knots

   (3) Pulse repetition rate, antenna beam pattern, pulse width, and other details are identified in Specification MIL-E-784K for AN/APQ-28 radar.

   (4) A device to simulate shadows, aspect angles, and reflectivity characteristics.

   c. Resolution of simulated radar—horizontal, 100 yards; vertical, 3 yards.

   d. Size of simulated area, 1,500 by 800 nautical miles.

   e. Accuracies of simulated radar.

   (1) Bearing   ± 2 percent

   (2) Range   ± 5 percent

   f. Size of simulation system, 7 feet high by 16 feet wide by 4 feet deep.

   g. Weight, no restriction.

   h. Flexibility—design should be sufficiently flexible to permit modifications of simulated terrain features and radar characteristics by qualified maintenance personnel in the field without excessive equipment down time.

   i. Testing of simulator—in accordance with test procedures documents as approved by the contracting officer.

   (1) Environmental—in accordance with Specification MIL-T-17113.

   (2) Performance—as necessary to certify compliance with Simulator Specification N1001. Detailed tests are to be established by the procuring activity within 100 weeks of the contract award date.

   j. Simulator reliability equipment.

   (1) Specified MTBF (mean time between failures), one hundred (100) hours.

   (2) Specified MTR (mean time to repair).

   (3) Continuous operation capability, twenty-four (24) hours.

   k. Government-furnished data is to be provided on landmass details.

3. Required technical and administrative monitoring items (Appendix A).
   a. Engineering reports.
   b. Programming or transparency system reports.
   c. Engineering drawings.
   d. Software development documents (if computer system used).
   e. PERT reports.
   f. Schedule prediction reports.
4. Required support documentation (Appendix B).
   a. Installation manuals.
   b. Maintenance manuals.
   c. Special tools list (software and/or hardware).
   d. Spare parts list.
5. Delivery of trainer, one hundred and thirty (130) weeks.
6. Referenced general specifications.

MIL-T-17113	Tests, shock, vibration, and inclination (for electronic equipment), general specification.
MIL-E-16400	Electronic equipment naval ship and shore, general specification.
MIL-E-784K	Radar set AN/APQ-28, general specification.
AIR-3111-962	Standards for engineering design reports.
AIR-423-912	Standards for engineering drawings.
AIR-512-352	Requirements for maintenance manuals.
AIR-3102-100	Military training devices, general specification.
MIL-STD-1690	Computer programming standards.

A separate summary reflecting the special points cited in the statement of work (SOW) which are not covered in the specification summary should be prepared and used by the project manager to crystalize and make constantly available other salient requirements of the procurement contract not specifically covered or detailed in the specification.

2.5   Analysis of Requirements

The specification for the RLM simulator procurement constitutes a modified performance specification, since it overtly restricts the acceptable design approaches to either the electro-optical system (using transparencies to store terrain data) or a computer system (using software for such data). Occasionally, an apparently pure performance specification imposes limitations on the design approaches that can be used through some restrictive language. For instance, the specification cited herein could have required an approach based on utilizing special transparencies of the areas to be simulated that would be provided as government-furnished equipment (GFE). Such a specification requirement would have eliminated feasible design approaches that could have met the performance requirements without utilizing the GFE that was mandated for use in the simulator.

Other types of specification language could inadvertantly (or otherwise) restrict the design options of a performance specification. When restrictive language is noted in a procurement specification or other document which works to the disadvantage of the offeror, it is the project manager's responsiblity to question the procuring authorities about this language and seek a clarification or revision.

Throughout the program, the project manager must keep in continuous focus the major requirements so as to avoid having irrelevant factors influence a technical or administrative decision.

After carefully studying the specification and all referenced documents and having satisfied themselves that they fully understand the specification requirements, the project manager and the staff must select the technical approach which best satisfies the following criteria.

1. It offers the best design for the equipment under procurement and is consistent with the specification requirements.

2. It is best suited to the experience, facilities, and capabilities of the company.

3. It will permit the company to offer the lowest cost and best delivery for the procurement.

In some cases, no one approach would satisfy all of these requirements. When such a situation occurs, a carefully thought-out compromise must be selected that offers the best chance of obtaining the contract, meeting the contract requirements, and making a profit.

In addition to the major procurement item, which is the equipment, the project manager must consider and plan for the other items of the procurement. One important tool used by managers for crystallizing the scope of a project or any element is the work breakdown structure (WBS). A WBS for the procurement project includes the major line items of the schedule (such as the equipment itself, data, and support items) and provides a means for identifying the composition of each item. Figure 2.1 illustrates a partial breakdown for a computer project. Each element can be broken down into as many tiers of details as are necessary. The WBS is used as a basis for cost estimating, categorizing types of effort, deriving resource requirements, scheduling, and deriving other planning information.

2.6 Selecting the Design Approach

Since the RLM simulator described in the specification document of the RFP indicates two acceptable design approaches, the objective analysis of each different design approach enables the project manager to converge on a decision regarding the one design that best satisfies the criteria discussed in Section 2.5.

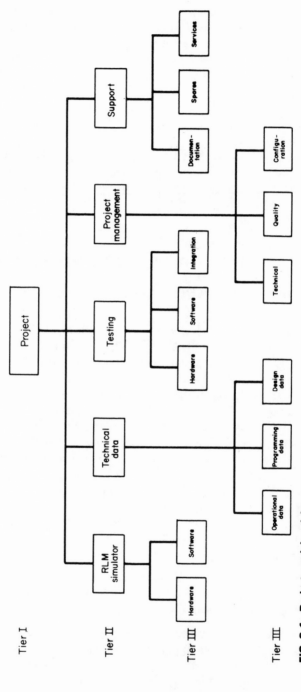

FIG. 2.1 Project work breakdown structure.

The two technical designs that are permitted are as follows: (1) an electro-optical design utilizing transparencies that store the terrain characteristics of the gaming area, and (2) a digital computer system in which the terrain characteristics are stored in computer programs.

To illustrate the analytical processes in which the project manager engages in deciding on a technical approach for the procurement in question, a basic description of each design is given with an analysis of how well each approach meets the specifications, the company capabilities, and the cost objectives. In actual practice, the project manager is already familiar with each design or is required to obtain the information from available sources of this knowledge.

In the electro-optical system, the data relating to the terrain characteristics are stored as coded information on two aligned transparencies. Electronic beams from flying spot scanners illuminate the transparency areas that relate to the exact comparable area being probed by the radar of the simulated training mission. Detectors sense the scanner light, which has been modulated by the transparencies, and the resulting signals are processed to provide a duplication of the radar presentation that an operator would witness in an actual mission.

One of the many points to consider is that the electro-optical system would present difficulties in meeting the flexibility requirements of the specification as far as changing the terrain characteristics is concerned. The transparencies which store the terrain data are planned for a 5,000,000:1 scale and require highly skilled craftsmen and equipment for their creation. Revised terrain requirements would entail expensive and time-consuming efforts by the transparency craftsmen which would not easily comply with the requirement concerning the capability to modify the simulated terrain characteristics "without excessive equipment downtime." If the scale were reduced to a more convenient size, other design parameters of the simulator would require revisions and would present serious problems that would render that system design difficult or unfeasible from other points of view, such as physical size.

It should be noted that it is very important for the project manager to solicit information that defines "excessive downtime" from the procuring activity, since the decision relating to the use of the dual transparency system may well be resolved on the basis of that point alone.

The digital computer approach is the second acceptable method that must be considered. Instead of storing the minute and detailed terrain data on a transparency, the terrain data are programmed for the digital computer. The geographic area of 1,500 by 800 miles must be divided into small area sections of discrete points, each point having its individual characteristic that affects radar returns, such as elevation, type of topography, type and configuration of objects or structures, terrain features, and other information that is stored in the memory of the digital computer. The maneuvers of the simulated aircraft and the operation of the radar antenna generate coded signals that identify the area sections illuminated by the simulated radar beam. The terrain characteristics, relative location of the aircraft and radar beam, and other pertinent information are processed by the digital computer system in real time converted to video information and displayed as a dynamic radar landmass to the student radar operator.

Since the computer system design for the RLM simulator represents a relatively new approach, a more thorough analysis of the specification requirements is required by individuals experienced in the field of computer sciences. Determinations must be made to establish whether the quantity of data to be handled, the computational speeds that are required, and the accuracy of the outputs can be achieved by utilizing any of the computer systems that are available on the market. In addition, the analysis must establish whether the other specification requirements (such as physical size, as noted in Section 2.4) would be satisfied by the digital computer approach.

It should be noted that rapid advances in technology continue to occur over the broad front. Such innovations, including state-of-the-art advances, are particularly prevalent in the field of computer sciences. Unachievable objectives of the recent past can often be routinely handled with a new-generation computer system. In addition, innovative mathematics modeling and programming techniques have been instrumental in making feasible the use of digital computers in many new applications. The procuring activity in our example had determined that an offeror expert in the computer science field could, in all probability, develop an acceptable system based on the computer technology and thus realize the inherent advantages of reliability and flexibility for the RLM simulator. Therefore, the specification and other procurement documents were written to consider the computer design approach as well as the established electrooptical system. A company such as Creative Electronics Corporation would be remiss in offering the proven transparency system without considering the many potential technical and cost advantages that might be available with the computer system design.

Before a particular technical approach is established as a candidate for a system design, the project manager must determine whether the performance objectives of the equipment can be achieved using the technology under consideration. For the simulator, the basic requirements relating to the system's ability to store the required data and the computational speed necessary to process data to permit its display in real time are prime considerations regarding the possible use of a computer system.

For the simulator, discrete points representing the terrain characteristics of the 1,500 by 200 mile area must be stored in memory. Since the resolution requirements permit the area to be divided into a series of grid lines 100 years apart, each intersection of the grid lines would represent a terrain point to be stored, and the terrain characteristics that must be simulated would be included in each point as a coded computer word. Based on the above-noted requirements, the computer must have the capability to store 4.8×10^8 simulated terrain points.

In addition, a review of computer capabilities is required to determine whether the other performance requirements of the specification can be met with a digital computer system. A determination must be made as to whether computers can process data fast enough to provide a dynamic display in real time. The analysis would reveal that in any particular instance only a small portion of the total terrain data stored would require processing, since in an actual mission, the radar beam would be restricted to a small portion of the total mission area. The establishment of this type of information by an effective analysis of the specification requirements by the project management team can often point the way to a unique design approach that can provide a significant edge over competing offerors. In the case of the simulator, a design approach based on handling only a limited area instead of the total gaming area of 1,500 by 800 nautical miles greatly relieves the computational speed requirements of the computer system, since only 2,000 square miles or 8.0×10^5 grid points must be processed in real time instead of 4.8×10^8 points.

When the basic memory, computational speed, accuracy, and other performance criteria that are necessary for meeting the RLM simulator requirements are established for a digital computer approach, the project manager must determine whether any available processors are capable of satisfying the project specification. The analysis calculations, based on an array of the latest generation of computer processors and programming approaches, can demonstrate that the digital computer approach is technically feasible for the RLM simulator application.

After considering the relative technical feasibility of the permitted design approaches, the project manager must determine which approach is better suited to the resources, experience, and facilities of the company. The following factors require consideration.

1. The approach that is compatible with the experience, facilties, resources, and capabilities of the company: The transparency approach requires personnel highly skilled in optics engineering, photographic technology, and electronic design. The digital approach requires computer scientists, programmers, and electronic engineers. Whereas the Creative Electronics Corporation has a background and capability as an optical and photographic concern, the company has recognized the potential of computer systems and has developed an effective in-house or subcontractor capability in that area. The analysis by the project manager would conclude that the Creative Electronics Corporation has basic experience, facilities, and capabilities adequate to offering either the transparency or the computer design for the procurement.

2. The approach that offers the better design for the system: Each of the two design approaches under consideration involves efforts that are unique. The

transparency system requires competent resources in the areas of optical and electronic engineering. In addition, the creation and modification of the transparencies that are used as the data base for the radar displays requires special equipment and resources.

The computer approach involves the use of computer scientists and electronic engineers. The creation of and modifications to the data base would be accomplished by computer programmers, but no special equipment would be required for that specific task.

A consideration of the various factors noted above and the analysis of the cost and the company's effectiveness in providing field support for the equipment and the requirement for revising the transparency data base over a 2-year period indicates that it would be to the advantage of Creative Electronics to propose the computerized RLM simulator instead of the electro-optical design.

It should be noted that the flexibility requirement and other ambiguous points of the specification were quantified by the procuring agency as a result of questions asked by representatives of offerors during clarification meetings.

3. The approach that will permit the company to offer the lower cost and better delivery for the equipment: In a line organization, the project manager would derive the cost and delivery estimates and reach conclusions on the basis of work breakdown structures and similar analytical tools. The project manager in a matrix organization would normally utilize information provided in work packages and would converge on conclusions resulting from discussions with the functional manager.

The comparisons of the analysis would reveal that the cost and time the different types of effort would require for the transparency and the computer design approaches would tend to compensate each other. The programming effort, for instance, would require the same number of man-hours as the effort to create the plates requires in the transparency system.

4. The approach that presents the lesser risk to the company: Any developmental work entails a degree of risk, which must be assumed by one or more parties to the project. Since the procurement case under discussion requires that the contractor assume the risks for the effort, the magnitude and scope of the risk must be evaluated and reflected in the proposal price and delivery commitment. Since the Creative Electronics Corporation has the capability, experience, facilities, and resources for either design approach, and since the required technological development is not beyond the corporation's capabilities, the only type of risk for either system would be those related to ineffective management, engineering and programming errors, inadequate testing, etc.

To conclude, the analysis would lead the project manager to choose the computer system on the basis of the following conclusions.

1. The use of the computer system approach is technically feasible for application.

2. All specification details, including the flexibility requirement, can be readily satisfied only with the computer system.

3. Creative Electronics has the background resources, facilities, and know-how necessary for the computer approach.

4. The computer approach can be implemented at a competitive cost and within the delivery schedule that is specified.

5. The company would not be assuming any unusual risk by offering the computer system.

2.7 Summary

The primary documents comprising a procurement package are the covering letter, the SOW, the specification, the contract schedule, the TPR, the cost breakdown, and general and administrative clauses and data. The SOW contains information that supplements the specification and serves to provide the type of information that assists companies in their decisions as to whether or not to participate in the procurement.

The project manager must consider many factors prior to determining whether the company should expend the proposal effort for a procurement. The major factors for consideration are availability of technical and other resources, facilities, work load, competition, delivery schedules, and risks. The WBS is often used to identify the score of the procurement and therefore to assist in the decision-making process. The process of arriving at the decisions is different for managers in organizations than for those in line matrix organizations.

The specification is the key document for a procurement and should be summarized and continuously referenced by the project manager. A performance specification addresses the results that are required of the equipment, leaving open or limiting the options as concerns the design approach to be used.

The project manager must analyze the specification to establish which design is best suited to the experience and resources of the company after the decision to participate in the procurement is reached. Risk is a major factor that must be objectively evaluated.

BIBLIOGRAPHY

Defense Contract Management for Technical Personnel, Naval Materiel Command, Washington, D.C., 1980.

Planning and Control Techniques (AMCR 11-16), vol. 3, U.S. Army Materiel Command, Washington, D.C., 1963.

3

DESIGN PROCEDURES
AND OBJECTIVES

3.1 Preliminary Design Approach Considerations

The customer's evaluation of the design approach, as described in the technical proposal, represents the major factor that will determine whether or not an offeror will be successful in winning a contract award. In many types of negotiable procurements, if the proposal contents (such as delivery and price) are marginally unacceptable, revisions can be negotiated. In addition, if the technical approach that is offered contains some flaws which are not inherent in the design, revisions by the offeror are normally permitted. However, a proposed design approach that clearly cannot satisfy the requirements because of basic design shortcomings is judged as not being susceptible to being made acceptable. Technical discussions of design-approach inadequacies generally relate to ambiguous requirements that were incorrectly interpreted by the offeror, inconsistencies, omissions, or other defects in the procuring agency's solicitation documents. Therefore, during the solicitation or contract definition phase, the primary concern of the project manager is to make sure that the design approach that is described is technically sound and is responsive to every point cited in the technical proposal requirements (TPR) documents. Since the details of the TPR are derived directly from the specification, the design approach that is proposed is the one that would eventually be used by the successful offeror in making the equipment. To make sure that the design that is proposed is the one that not only is fully responsive to the proposal requirements but also involves the technology with which the offeror is competent and experienced, a comprehensive technical analysis should be conducted.

The technical analysis required for the development of a computer system such as the RLM simulator unfolds in the four basic phases depicted in Figure 3.1: (1) system engineering, (2) software and hardware design, (3) program writing and production, and (4) integration testing.

During the development of complex equipment, the system engineering process is conducted several times. Each effort establishes more detailed and comprehensive designs and is used to implement necessary revisions. The product of the system engineering effort includes detailed descriptions of modules, the identification of signals, interface data, and other elements that must be handled by the equipment software and hardware.

The hardware design consists of translating the functions to be performed by the hardware circuits into drawings, production specifications, and other types of data that would be used for the production, fabrication, and testing of the hardware. The software design establishes computer performance capabilities, programming and coding requirements for each module, and details of how the computer program will be structured to satisfy the requirements.

In order to provide visibility for the creation of software, rigorous controls such as the configuration management disciplines are applied. One result is that the vague line of demarcation between software design and production phases that was characteristic of earlier software efforts no longer exists on contemporary projects. The writing of the program constitutes the production phase of software creation. Comprehensive and accurate documentation of the computer program design must be approved prior to the start of the program writing effort.

The production of the hardware is initiated by the approval and release of production orders, work packages, or other instruments to the production manager for implementation.

For software, the production or program-writing phase is initiated after approval of documents that provide detailed design information concerning

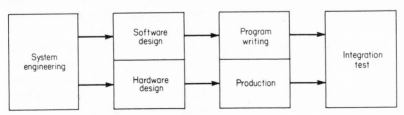

FIG. 3.1   Computer system formulation.

function, interface, timing, logic, language requirement, and other data relating to each of the modules for which a program will be written.

The integration test portion of the system formulation effort involves providing tests and whatever corrective actions are necessary to make sure that the hardware and software designs are compatible and that the signals that flow among the system modules provide the specified results.

3.2 Systems Engineering

In Figure 3.2 the three divisions of effort which translate the performance requirements into engineering criteria are identified as function analysis, system breakout, and system quantitation.

The primary object of the function analysis effort is to express the specification performance requirements in functional and engineering terms. The translation of those performance requirements requires the creation of the mathematical models for the various systems functions and the determination as to which functions are to be performed by the software and which by the hardware process. During the analysis, functional block diagrams are created which provide information regarding the relationship of the various hardware and software actions, the interfaces that are required, and, in general, an overall engineering view of the complete system.

Other products created by the function analysis effort include reports and data describing the various functions of the block diagrams. The reports also address other specification requirements, such as reliability and flexibility, which cannot be depicted in the block diagram format.

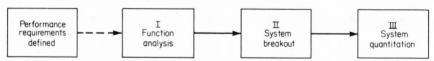

FIG. 3.2   System engineering categories.

The data established during the function analysis category of the systems engineering phase is expanded by the system breakout effort identified as category II in Figure 3.2. The data that is generated includes schematic block diagrams and a more detailed description of the functions among the different hardware and software elements of the system. Interfaces that are required to provide compatibility and to facilitate signal flow among the various subsystems and modules are identified and described. It is during the system breakout portion of the systems engineering effort that the equipment work breakdown structure (WBS) for the hardware and software is developed.

Systems quantitation represents the third category of systems engineering. During this portion of the effort, the previously established data are numerically defined and expanded in other terms, such as equations that can be established as design or program criteria. The types of engineering data that are derived include input-output functions among computers and other system elements, rates of data processing, data formats, mathematical expressions to be mechanized or programmed, data execution times, memory and accuracy requirements, and other similar information. Of particular importance to a computer system is the necessity of providing for adequate memory and I/O (input-output) capacity. Experience has indicated that during the development of new computer systems, demands for memory escalates and shortages in already selected computer components tend to handicap the efforts of the project. Since the initial cost of providing for spare memory is not significant, the project manager should direct that at least 50 percent spare memory capacity be provided in the design.

Providing spare I/O capabilities is also essential, since the computer program may be required to function with additional lines in the future. Thus the practice of providing some 25 percent spare I/O capacity represents prudent planning of computer systems.

The information established by the system quantitation effort serves as the basis for selecting the general purpose computer that would be required and provides for the specific WBS of the computer program. The effort also creates information that represents the functional parameters of the hardware that would be designed and fabricated for the system.

The quantitation effort results in an array of different types of data that serves as the basis for the design of the equipment systems and computer programs. The types of documented data include the following.

1. Hardware design specification
2. Program performance and design specification
3. Data base design documents
4. Software functions, numerical values, I/O parameters, and other mathematical data
5. Program flow charts

During the course of a development, the project manager has different categories of the systems engineering effort repeated in greater depth several times to provide the degree of detail that might be required at any point in a development project. For example, the technical proposal requirements usually require a greater emphasis on obtaining data from the function analysis category than from the system quantitation category of the system engineering work. If a contract is received, the categories of the system engineering effort are repeated to provide information with far more emphasis on the system quantitation calculation so as to obtain the type of criteria required to implement the design of the equipment and computer program.

The actual creation and management of the computer program require a rigorous monitoring, testing, documentation, and reporting effort that will be identified in subsequent chapters. The main point that the program manager must keep in mind is that the program effort generally constitutes the major expenditure of resources, money, and time. In addition, the greatest problems, as far as meeting the schedule, performance, and cost objectives of a project are concerned, are due to programming difficulties. Therefore a comprehensive analysis of the software requirements will enhance the manager's ability to prepare proposals that are realistic.

3.3 Equipment Design Approach

The implementation of the system engineering effort identified in Figure 3.1 and the objective consideration of the various technical and commercial factors represent one of the most important steps in the execution of a development project. The design approach that is established by the system engineering effort and described in the technical proposal constitutes a major factor in determining the contract award.

The success that the new contractor experiences in meeting the performance, schedule, and project objectives of the contract is based on the excellence of the system derived from the system engineering effort.

In the case of the RLM simulator, the ultimate responsibility of the project manager is to make sure that the design approach is the one that most effectively meets the performance requirements of the procurement and is compatible with the company's available resources, technical capabilities, and experience. The simulator system depicted in Figure 3.3 serves as the case project. An overall description of the major functions of each of the blocks shown in Figure 3.3 will serve to permit an appreciation of the role that the project manager plays in directing the engineering tasks required in developing the system.

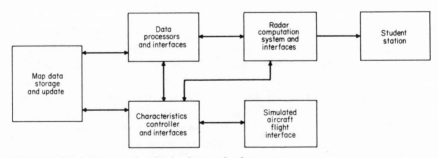

FIG. 3.3 Block diagram of radar landmass simulator.

The functions performed by the map data storage subsystems noted in Figure 3.3 provide the basic terrain data that includes the characteristics of elevation, reflectivity, cultural data, and other types of information necessary for displaying realistic radar returns. In addition, the subsystem provides for the selection of the information to be displayed on the basis of the dynamic movement of the simulated aircraft and radar antenna sweeps.

The data processor subsystem performs the basic function of storing the data from the map data storage unit that is to be displayed at a particular point in time.

The characteristics controller provides the timing and sequencing functions for simulated aircraft motion signals and the stored data. In general, the subsystem performs an overall signal coordinating function for the different signals in the simulator system.

The radar computational subsystem solves, in real time, the many equations relating to each point to be displayed.

The foregoing material constitutes a very brief description of a very complex computer system. In effect, the RLM simulator system must be capable of providing simulated flights in any direction over the gaming area, with the radar antenna sweeping the terrain and displaying the terrain characteristics illuminated by the radar on the student display. To accomplish these objectives, the various subsystems shown in the block diagram of Figure 3.3 must make calculations in real time of such parameters as shadow effects, Earth's curvature, aircraft elevation, variations in radar returns due to different characteristics of the area of interest, and slant ranges. The project manager is responsible for directing and coordinating the efforts of the project team in the designing, programming, testing, manufacturing, etc. of this complex system.

In addition to the block diagram, the WBS is a key product of the systems engineering effort. The overall WBS for the project, including the major procurement items, is presented in Figure 2.1. The portion of the structure relating to the RLM simulator represents one portion of the WBS that is expanded as Figure 3.4 and is derived as follows:

1. Translation of the performance objectives into hardware and software subsystem elements

2. Division of the subsystem elements into tiers of design

3. Selection of feasible subsystem designs that are compatible with each other and will satisfy the specification requirements

4. Review and revision of the system design approach so as to reach the optimum combination of cost, performance, and schedule objectives

In developing the design approach to be proposed, the project manager is interested in the tier of detail that provides general intelligence relating to the hardware to be fabricated or purchased, the identity of the software that will be required, and the general scope of effort that will be necessary to create the

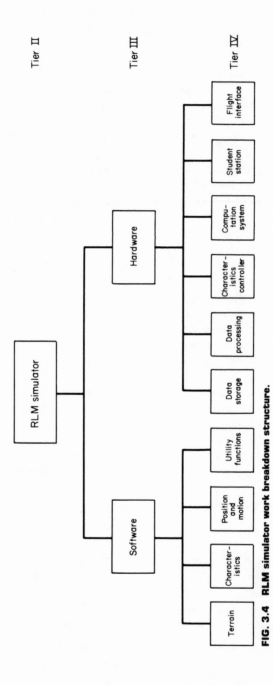

FIG. 3.4 RLM simulator work breakdown structure.

system. The design of a WBS for a particular system will vary among project managers or engineers, since the breakdown information is a reflection of how the creating individual perceives the structure and of how the project will be implemented. Thus, as a tool, the WBS must serve the particular project team and must be tailored accordingly.

Figure 3.4 identifies the areas of effort required to create the simulator under contract. As noted, the creation of the software and the creation of the hardware represent the two major categories of effort. The different elements making up the tier IV breakdown will show a general similarity to the various subsystems that will describe the system block diagram to be developed as part of the analysis effort. The WBS is usually broken down into the degree of detail that the project manager judges best suited to the project effort. The final breakdown is into specific work packages assigned to individuals. Each work package contains a budget, a schedule, performance criteria, and other information relating to the required effort.

In the line organization, the project manager assigns to cognizant engineers of the proposal team the tasks of establishing the design approaches and of providing the engineering that may be required for the subsystems identified in the work breakdown structure. The products that each individual produces are generally in the form of reports, schematics, and other types of documents that the project manger specifies. In the matrix, the discipline supervisors or their counterparts on the team are assigned tasks described in the formal matrix work packages. The packages are standard organizational documents which call for specific technical, cost, delivery, and other information that company policy requires. In both types of organization, the project manager is responsible for collating and editing the material that is submitted to include or highlight what is required by the TPR in relation to the design of the system that is to be proposed.

3.4 Specification Terminology

In order to identify various requirements applicable to complex systems procured by the government, different terms were coined to be included as specification requirements for weapon systems and other complex projects. The terminology succinctly describes important requirements in what is called "ility" jargon and, if overlooked, can have a significant impact on systems under procurement. The definitions of the more common terms include the following.

1. *Compatibility*: The ability of various subsystems to operate as an integrated unit over the entire range of functions can present engineering problems. For instance, for computer systems, specially designed interfaces between processors that convert signals between two dissimilar systems require special design effort and testing to make sure that the full range of system capabilities

can be realized. In addition, computer systems require that interfaces be designed to convert signals between dissimilar units so that they function together over the full range of their performance objectives.

Compatibility requirements are stressed for new systems that must function with existing equipment. In the case of RLM simulator, the system under procurement must include whatever interfaces are required to function with the specified existing flight simulator.

2. *Commonality*:   The word "commonality" originated in the Department of Defense in a discussion concerning the justification for awarding a major procurement in 1963. It crystalizes a factor inherent in the utilization of a multitude of similar pieces of equipment—e.g., the United States Navy's use of squadrons of identical aircraft.

Commonality is a factor which reflects the percentage of the components, modules, and other elements of a piece of equipment which are interchangeable with existing equipment of a different type. For instance, the Army may have in operational use a search radar using a particular cathode-ray tube (CRT) for the display of intelligence. For procurement of an entirely different type of radar, company A may propose using a CRT identical to that of the search radar in operation, whereas company B may propose using a completely new tube. Table 3.1 summarizes the various commonality factors and indicates why company A would be preferred to company B, all other factors being equal.

For a prototype procurement such as the RLM, commonality does not play any significant role. However, the project manager must be aware of this factor and must incorporate the maximum degree of commonality in any quantity procurement and describe how it is to be achieved in the technical proposal.

3. *Cost effectivity*:   Effectivity, as it relates to cost, is another new term that has arisen as a result of large Department of Defense procurements. The term relates to a concept that reflects the total cost to which a customer may be subjected if equipment of a particular design is purchased. To illustrate this point, consider the evaluation of two different designs of high-performance aircraft

TABLE 3.1   Summary of Commonality Factors

Commonality Factors	Company A	Company B
1. Savings due to elimination of second set of CRTs as spares	Favorable	Unfavorable
2. Advantages due to familiarity of maintenance personnel with existing CRTs	Favorable	Unfavorable
3. Interchangeability of CRTs with other radar systems	Favorable	Unfavorable
4. Savings due to quantity purchase program for CRTs	Favorable	Unfavorable

offered by two competitors. Assume that aircraft A costs $100,000 per unit more than aircraft B. Aircraft A can land and take off on the average 10,000-foot runway, whereas although aircraft B, although less costly, requires a 15,000-foot runway. The procuring agency making a study of runway lengths finds that the cost of increasing the length of the airport runway far exceeds the total cost differential between aircraft A and aircraft B. The evaluation of the total cost picture constitutes cost effectiveness. In the case of the two aircraft designs, the more expensive aircraft A would be the design to purchase.

The concept of cost effectiveness should be considered by the project manager in selecting a design approach for a procurement and should be discussed in detail in the proposal, particularly when the technical proposal requirements call for a cost-effectiveness discussion.

4. *Reliability*:   Reliability is the ability of a product to function as required without exceeding a specified failure rate. Reliability criteria are usually expressed in terms of MTBF (mean time between failures).

5. *Maintainability*:   This is the ability of a product to be repaired within a specified time. Criteria are usually expressed in terms of MTR (mean time to repair).

6. *Testability*:   This is the degree to which the structure and the content of the design of the hardware or software facilitate the step-by-step testing and verification of the capabilities of the product.

In addition, special terms for software have surfaced which relate to essential characteristics of computer programs.

Some of the major design objectives for a computer program include the following.

1. *Validity*:   Ability of a program to provide all the functions, performance, interface actions, and design capabilities without requiring supplementary manual or programming inputs

2. *Robustness*:   Ability of a program to recover and continue reasonable performance despite erroneous inputs

3. *Reliability*:   Ability of a program to perform all specified or documented functions without detectable errors

4. *Maintainability*:   The degree to which software documents are complete, accurate, and structured so as to permit revision and repairs with a minimum of delay or man-hour expenditure

5. *Testability*:   The degree to which the structure of program modules facilitates step-by-step testing of all functions

A well-designed computer program inherently possesses the foregoing features. However, it is incumbent upon the project manager to make sure that these features are reflected in the design.

3.5 Summary

Subsequent to the decision to participate in a procurement and the determination regarding the technology to be offered, the project manager must make an appraisal of the scope of the procurement project, establish the basic design approach which will satisfy the specification and which is compatible with the capability and resources of the offeror, and direct the procedures for the design of the system. A system engineering analysis establishing the design approach for the equipment is required for the proposal and for the engineering design for the equipment under contract.

The factors that affect the system approach are the specification requirements, company resources and experience, the cost delivery schedule, and risk factors. The project manager is responsible for considering and evaluating each factor prior to determining the design approach to be used.

Completion of the various phases of a system engineering analysis is essential to the derivation of the optimum design for the equipment. The first or functional analysis phase is to identify the major functions to be peformed by the system. Functional block diagrams and expanded WBSs for the equipment are essential tools in this phase. The system breakout or synthesis phase provides a greater depth for the design detail established during the first phase. The third, or system quantitation analysis, phase provides design data and calculations which establish the detail design approaches, accuracies, and other data needed to meet the specific requirements of the system.

For a computerized system such as the RLM simulator, a computer program analysis is required which establishes the preliminary and design specifications for the computer software. Since the creation and testing of the computer software and its integration with the hardware constitute a major area of cost and effort, the program manager must provide rigorous control and surveillance in completing the required programming effort.

Other requirements that are often included in the specification which must be considered in the equipment design approach are commonality, compatibility, cost effectivity, transferability, and similar equipment features that are commonly included in government specifications. In addition, new terms to express software qualities must be recognized and implemented in computer program designs.

BIBLIOGRAPHY

Summer, C. F., *Simulation Systems Programming Design Manual*, Naval Training Equipment Center, Orlando, Fla., 1973.

Fife, D. W., *Computer Science and Technology*, U.S. Department of Commerce, Washington, D.C., 1977.

PREPARATION
OF PROPOSALS

4.1   Technical Proposal Requirements

Because of the costs that offerors must incur in the preparation of proposals and because of the time and effort that procuring activities must expend in evaluating and negotiating proposals, the participating offerors are usually limited to organizations that are judged to be qualified for the procurement. However, for government procurement, a particular company which may not be judged qualified can appeal to the contracting authorities for the opportunity to participate.

In the case of the RLM simulator project, the project manager was successful in having the Creative Electronics Corporation judged as one of the companies qualified to participate in the procurement and was provided with the RFP (request for proposal) package. The TPR (technical proposal requirements) part of the RFP is the document that specifies the sequences in which the material is to be presented; the type of information required; accounting of hours that will be spent on engineering, programming, manufacturing, testing, and integration of the equipment; and other types of information which will be evaluated by the project team of the procuring organization. The degree of specifics that the TPR requires depends on how well the equipment design is defined in the specification. If the specification describes equipment that is to be designed using a particular technology, the TPR will require that detailed information be provided in the proposal for evaluation. However, if the technology to be used in the development of a desired system is left to the ingenuity of qualified offerors, as is the case for research- or development-type efforts, the proposal information that the TPR identifies will have to be general in nature. A hybrid specification is a performance specification which identifies two or more technologies that are acceptable for the design of the equipment to be developed or which imposes other limitations regarding the approach to be used to meet the performance requirements of the equipment.

The RLM simulator is typical of the type of project that uses a hybrid type of performance specification. The design approach to be used must be either the electro-optical or the digital computer design approach. The areas cited in the TPR which will be evaluated either must be general in nature so as to be applicable to either technology or must provide for alternative information relating to a particular area applicable to the type of system proposed. For instance, to evaluate the resolution capabilities of the computer system, the TPR would require specific information on the computer program; the computer characteristics, such as word length; and similar information. However, to evaluate the resolution capabilities of the electro-optical system, the TPR would require information on such factors as the transparency scale and map details. Such appropriate TPR information permits the evaluator to judge the relative capabilities of the computer and electro-optical system for meeting the resolution requirements of the specification.

In order to facilitate the comparative evaluation of proposals submitted by different offerors, the content and format for the submissions are specified, and separate volumes are usually required for each of the following areas.

1. *Engineering presentation*: system design approaches and system architecture

2. *Implementation plan*: personnel, facilities, and scheduling

3. *Logistics support*: support documentation, spare parts, maintenance, and installation

When dictated by size, complexity, technology, or other factors, additional proposal volumes are required to evaluate specific other areas of computer software or special technology subject matter.

Since the technical proposals will be used to establish which of the systems described by the different participants in the procurement are relatively superior, the project manager's main concern is to direct the preparation of the best possible proposal. The areas that the TPR identifies for discussion in the proposals usually include the following.

1. Areas which are critical to the successful design of the equipment. In the case of the RLM simulator, such an area might be the method of storing and

processing terrain data, which is critical to meeting the resolution and accuracy requirements of the specification.

2. Areas which by their nature pose difficult engineering problems that must be successfully resolved—e.g., the method by which the characteristics (accuracies, beam pattern, etc.) will be simulated.

3. Areas for which major cost and effort expenditures are required—e.g., a presentation of the techniques to be used in storing and processing the terrain data for the simulated radar.

For government programs, the procurement regulations require that the relative importance of each area of the TPR be indicated and that the general criteria for evaluation be provided for all offerors. It is therefore very important that the offeror's project manager study the TPR document in conjunction with the specification and the SOW and direct the technical proposal precisely as required.

As part of the effort for organizing the proposal write-ups, a summary of the information required by the different volumes of the TPR should be documented by the project manager and utilized as a control to make sure that the material prepared by different members of the proposal team addresses the required points. Since the Creative Electronics Corporation will describe a computerized system in its proposal, the summary prepared by the project manager may have to translate the generic points of the TPR to make sure that appropriate computer programming information is provided, since the RLM simulator project does not specify that a computerized system must be provided. The programming information is necessary in order to describe how different performance objectives will be achieved.

The following is a summary of the TPR prepared by the Creative Electronics Corporation project manager that would be appropriate for the engineering presentation volume of the RLM simulator proposal.

SUMMARY OF TECHNICAL PROPOSAL REQUIREMENTS
RADAR LANDMASS SIMULATOR—DEVICE 5A1

Volume I. Engineering Presentation
   1. Introduction.
      a. Technical rationale for proposed design.
      b. Credentials of Creative Engineering relating to RLM simulator project.
      c. Special and overall technical features of proposed system.
      d. Rationale for choice of computerized approach over alternate electro-optical system permitted by specification.
   2. Technical Discussion.
      a. Design approach.
         (1) Functions and signal flow of each of the major subsystems of the simulator as presented in system block diagram.
         (2) Discussion of computer architectures, interfaces, and data handling media.

b. Real-time capability.

(1) Provide analysis of the computer system speed capability versus the system speed that is required for real-time presentation.

(2) Presentation of mathematical model relating to data processing and number and rates of inputs and outputs.

(3) Presentation of mass-storage transfer rate capabilities.

c. Resolution.

(1) Description of mathematical model and software designed to provide 100 yards horizontal and 3 yards vertical resolution.

(2) Description of processing circuits related to resolution requirements.

d. Accuracies.

(1) Description of mathematical model and software designed to achieve bearing and range accuracies of ±2 percent and ±5 percent, respectively.

(2) Description of computational techniques for processing information to satisfy parameters.

(3) Provide functional block diagrams to support narrative.

e. Simulator radar systems.

(1) Provide block diagram and description of the radar system.

(2) Provide computational data and processing system to describe simulation of AN/APD 28 radar and antenna characteristics.

f. Flexibility.

(1) Describe system features which readily accommodate changes made in the field arising from on-site assessments of terrain features.

(2) Present procedures and techniques for revising terrain data when terrain characteristics require revision.

g. Test program: Description showing how specification requirements will be met. The narrative is to cover:

(1) Hardware.

(2) Software.

(3) Integration of hardware and software.

3. Hour and material breakdown: For each major subsystem of the RLM simulator, provide the breakdown of engineering hours, manufacturing hours, and material costs.

4. Scheduling: Provide in a Gantt chart form the scheduled effort for each of the line items of the contract.

5. Experience and facilities: Provide a summary of project personnel experience and credentials.

4.2   Evaluation Factors

A project manager must be constantly sensitive to the thinking and desires of the project chief of the procuring activity. Whereas this text is concerned with the role of the project manager vis-à-vis the offeror and ultimately the contractor, consideration of some of the functions of the personnel representing the procuring activity is in order.

One of the prime functions of the procuring activity's project chief is to head up the evaluation team, which is responsible for judging which offeror is most qualified. (In order to avoid confusion, the text will refer to the head of the project team of the procuring activity as "chief" and to the head of the offeror's or contractor's project team as "manager.")

In addition to the TPR, the procuring activity team establishes the TEP (technical evaluation plan), which contains the criteria to be used for evaluating the technical proposals, the scoring plan, and the evaluation procedures. The formal TEP is mandatory for government procurements which involve developmental effort, since basic information relating to the proposal evaluation factors must be provided to all organizations submitting proposals. The identity of the evaluation factors and subfactors assists the offerors in understanding more precisely what information the procuring activity is seeking in the proposal.

In addition to the evaluation factors, information relating to the relative importance of the different technical areas that will be evaluated and scored is provided to the offerors in the TPR package. The specific scoring weights are kept in confidence, but astute offerors competing for a procurement can often deduce much of this important information, which might give them an edge on other offerors.

For the RLM simulator procurement, it might be assumed that the project manager of Creative Electronics, through discussions with the procuring activity engineers, using activity personnel, and through information obtained by the marketing force, was able to obtain important background information relating to previous experiences with electro-optical and computerized landmass simulators. The information is not only valuable for its influence on the engineering approach to be proposed, but it serves as a powerful argument for the proposed system when expressed in appropriate portions of the proposal. For example, the difficulties that using activities have experienced with electro-optical RLM simulator designs that have been delivered in the past and the technical difficulties that plagued earlier attempts to develop the simulator on the basis of computer technology represent valuable information for the project manager of Creative Electronics or any other offeror. The main points regarding the two alternative approaches that constitute marketing intelligence of interest to project managers can be summarized as follows.

1. Electro-optical trainer designs.
   a. Constraints on physical size and scaling (5,000,000:1) of the transparencies upon which terrain data is stored limit size of simulated gaming area, resolutions, accuracies, and other capabilities of the trainer.
   b. The revision of the data base contained in the transparencies that may be required for effective training requires expensive time-consuming effort by uniquely trained craftsmen using special tools. Therefore the electro-optical design is relatively inflexible for providing updated radar intelligence on existing areas or new simulated areas.
   c. The operation, maintenance, and logistic support of the electro-optical trainer design requires uniquely trained personnel and special tools for pro-

viding the type of calibration and servicing necessary to ensure effective utilization.

2. Computerized trainer designs.

a. The vast amounts of terrain data needed to provide the required resolution accuracies, cultural characteristics, and other information to be programmed and stored in the computer system have presented problems in the past.

b. The system must be capable of functioning at very high speeds in order to read out, process, and display in real time the particular data relative to the training mission. The computational speeds of previous generations of computers have presented difficulties in meeting the real-time requirements of the simulators.

The project manager, who is aware of the past experiences with the application of the two technologies to the design of RLM simulators and who is also informed of the latest developments in the two types of designs, would arrive at the following conclusions.

1. The transparencies used in electro-optical designs for RLM simulators remain limited in terms of their capabilities for storing large amounts of data, their ability to provide more stringent resolutions, and their capacity to provide accuracies that will accommodate the design of trainers to be used for the advanced radar systems.

2. There have been no innovations which permit the creation of transparencies for specific geographic areas; they are still difficult to create and are even more difficult to revise when the circumstances require the updating of information for training.

3. Advances in computer technology have occurred over recent years, so that modern systems are capable of storing the large amounts of terrain data that previously represented a limitation on their use for simulators.

4. Innovations in integrated circuit technology that have taken place in the recent past have provided computational speed capability adequate to meet the demand for real-time data processing required of the RLM simulator design.

5. A computer system for the RLM simulator can be designed to store, off-line, all the basic data that represents any desired area. A computer system can be proposed that uses a control system that selectively utilizes only the specific area transversed by the simulated aircraft and the sector being scanned by the radar antenna. The technique involves processing of only that small portion of the total array of data that is being utilized for the radar display, thereby enhancing the ability to provide real-time presentation.

6. Revisions to the terrain characteristics that may be required at the training site can be readily made by modifying the program. In addition, programming modifications can be made by qualified maintenance personnel. The flexibility

requirement of the specification can be readily realized with the computerized RLM simulator design.

In organizing and editing the technical proposal, the project manager should continually focus on the following major objectives.

1. Making sure that every element cited in the TPR is addressed in the proposal.
2. Emphasizing the critical areas that the TPR identifies to provide comprehensive and crisp coverage.
3. Providing for the theme throughout the proposal that serves to emphasize points relating to the inadequacies of the electro-optical system and the advantages that would be realized through the new and innovative approach of using a modern computer system.
4. Avoiding irrelevant material.

4.3 Proposal Organization and Language

The organization of each volume of the proposal should follow the format and sequence usually provided in the TPR. For the RLM simulator, the organization of the technical proposal should follow the material previously provided in the "Summary—Technical Proposal Requirements."

The design approach derived from the system engineering analysis constitutes the technical base from which the project manager initiates the organization and coordination of the proposal effort. Since the TPR identifies the specific areas that will be evaluated, the effectiveness of the proposal requires that the project manager direct the proposal write-up in such a way that the proposal clearly and comprehensively addresses the points cited in the TPR, as discussed in Section 4.2. For instance, the section of the technical proposal that addresses the area of the TPR identified as "real-time capabilities" requires contributions from mathematicians, computer scientists, and programmers, since the achievement of the objective of that section requires a successful coordinated effort from those parties. It is the responsibility of the project manager to make sure that the material from the contributing parties is edited, structured, and coordinated in a clear and logical way so that the evaluators of the procuring activity have no difficulty in concluding that the proposed system will meet the real-time requirements of the specification and, it is hoped, that the design approach that is offered is superior to those proposed by competitors.

In the line organization, the manner in which the proposal effort is implemented is generally left up to the project manager. The format for each of the contributors varies in each case.

The form and content of proposal material and procedures used in matrix organizations are more formal. Work packages for each area are structured with a standard format for technical content and man-hour requirements, facilities, material, scheduling, and other associated information. Although the project manager has the prime responsibility for creating the proposal, coordination with the functional manager is required because of the manpower, facility, and scheduling requirements.

The narrative of the technical proposal should be written in such a manner and style that it presents a clear and concise picture of what is being offered; the narrative should also be sufficiently varied in style to hold the interest of the reviewer. A certain amount of psychology can be fruitfully exercised here. For instance, if it is determined that the customer's evaluating personnel are relatively inexperienced in a particular technical area, the proposal could discuss some of the basic fundamentals of the subject as applied to the proposed design, thereby enhancing the reviewers' grasp of the approach and facilitating an appreciation of the advantages that the particular design offers. One of the worst things to do is to submit a technical proposal that makes what is being offered difficult to understand. A reviewer cannot be expected to recommend an approach that cannot be readily understood, regardless of its inherent merits.

4.4 Introduction of the Proposal

In the proposal introduction, the discussion must be confined to those qualifications that relate directly to the procurement. For instance, it might be an established fact that Creative Electronics Corporation enjoys a reputation for being expert in the field of accelerometers, but since this particular capability is unrelated to the procurement under consideration, the impact of the proposal would be dissipated if the proposal discussed this capability of the company.

The following illustrates some of the points to be included in the introduction section for the RLM simulator proposal:

1. Statement of qualification.
   a. Accomplishments of Creative Electronics Corporation in developing and designing computer and electronics systems.
   b. How the accomplishments cited above qualify the company as competent to design the various subsystems of the RLM simulator.
2. Company technical credentials.
   a. Tabulation of government and commercial contracts and orders in areas related to the RLM simulator, giving a résumé of type of contract, delivery, and cost performance in each case.
   b. Summary of company-sponsored research programs and how they might relate to the procurement.
   c. Summary of any distinctive achievements, especially if the area of achievement is related to some facet of the procurement under discussion.
   d. Review of technical problem areas that were encountered in the past

and how the company, by creative engineering and ingenuity, was able to solve the problems.

e. Review of programs that were successful in past procurements because of sound and simple equipment designs.

3. Statement of broad design concepts.

a. This part of the introduction should serve as a summary of what is to be offered, describing the design approach proposed.

b. A discussion of how and why the computer system being offered incorporates advanced technology that will provide equipment performance capabilities that were not available in past computer systems and are not available in existing electro-optical systems.

4.5 Technical Discussion

The TPR provides the identity of the topic to be discussed in the proposal and also reflects what the evaluators for the procuring activity have established as the essential parameters of the equipment under procurement. In preparing the proposal technical material, the project manager should recognize that the evaluators will be under schedule constraints in their efforts to establish the relative merits of several different proposals. The evaluation team will have neither the time nor the inclination to make interpretations of or draw conclusions from otherwise ambiguous material. Most evaluators use scoring techniques for translating what is provided in the various sections of each proposal into a form that can be used to compare the responsiveness of different proposals to the TPR criteria.

Therefore, the project manager should strive to clearly and comprehensively address each point in the TPR to eliminate or at least minimize areas that must be clarified for the evaluators. The task of the evaluators for the procuring activity would be greatly simplified if one of the technical proposals could be judged as clearly superior to all others that were submitted. The objective of an offeror is to be successful in having a technical proposal prepared that will receive an outstanding evaluation.

The basic information contained in the "design approach" portion of the technical proposal is derived from the engineering analysis that was completed as a preproposal effort. The technical discussion that addresses the design approach should provide a depth of data and information that will hopefully leave no further questions for the evaluator. For example, to satisfy a TPR requirement that the prime functions of the signals that flow among the major subsystems of the simulator be presented (see 2.a.(1) of Volume I of the TPR), the block diagram for the RLM simulator (Figure 3.2) would have to be expanded and discussed in greater detail in order to provide comprehensive information that would leave the evaluators with no further questions.

Another example of the required proposal information relates to the real-time capability of the simulator. The time required for the computer system to process different categories of information varies, depending on the complexity and number of variables that must be handled. The proposal should provide a detailed description of the sequence in which each of the simulator subsystems would process a series of the most complex terms of the mathematics description of the radar parameters, terrain characteristics, and associated cultural data. Sample calculations, tables, and issuing diagrams that enable the evaluator to readily follow the sequence of time required by the computer system to complete the most demanding calculations within the necessary time frame would be essential to obtaining a high score for that particular proposal section.

The digital computer system being proposed represents an innovative design with which evaluators may not be familiar. The inherent capabilities for meeting the specification requirements for accuracy, fidelity of simulation, flexibility, etc. are considered by the project manager and the proposal team of Creative Electronics to be superior to the more traditional RLM simulator designs. However, it is up to the project manager to translate the information on the superior characteristics of the computer system into convincing proposal language relating to the specific points mandated by the TPR.

Based on the type of contacts with user personnel mentioned in Section 4.2, flexibility can be deducted as being a simulator characteristic that is highly important to the user. Since the computer system being offered lends itself so effectively to satisfying this requirement, special attention should be given to making sure that the write-ups clearly provide the desired message. In particular, the type of date storage media to be used, the procedures and time required to revise terrain characteristics or change the gaming area, and similar features relating to the flexibility of the system should be detailed in the proposal. In addition, since the availability of trained manpower resources is a critical problem for military and other organizations, the proposal should stress that even though qualified computer programmers must be used to reprogram and service the RLM simulator, they are far more easily available than the rare and highly specialized craftsmen required to modify the content of the transparencies that must be used with the electro-optical system design.

In general, the effective proposal will faithfully follow the format of the TPR and comprehensively address the points that are to be covered in the proposal. A proposal that includes, in addition to the engineering and design approach information, other information that is relevant and is perceived to be important to the using activity could provide an edge over competing offers.

The test program for a development project is usually of interest to the procuring activity, since the information from the tests will provide insight into how well the equipment will satisfy the specification requirements. In the case of the RLM simulator, for example, the effectiveness of the training that the equipment will provide depends on the realism of the radar presentation as the simulator maneuvers in different directions over the target area. Therefore, the proposal must indicate the scope and depth of testing that will be implemented as the design progresses to assure the evaluators that critical performance objectives—such as real-time presentations, accuracies, fidelity, and other key characteristics—will be achieved by the modules and circuits that are designed.

In addition to the test criteria and procedures that are to be used for the hardware, a detailed description of the test program for the software is necessary to show that the quality of the computer program is acceptable. Some of the major points that must be presented to assure the proposal evaluators that the software test plans are adequate and will result in computer programs of acceptable quality are as follows.

1. Independent software test team to be used
2. Description of software test tools that are planned for the effort (e.g., analyzers, data generators, comparators)
3. Identification of the test levels that are planned
4. Schedule of effort and resources to be applied
5. Test plans and procedures to be used

In order to cover all the points of design and at the same time have the technical proposal arranged in accordance with the TPR, a clearly expressed cross-referencing system is essential. In some cases redundancy may be in order. The project manager should keep in mind that the proposal may be reviewed by several individuals, each assigned to a particular section. It can be assumed, for example, that an individual assigned to analyze only the section dealing with testing of the simulator will read only that assigned section, even though a particular proposal may contain information on testing in other sections that may be germane to a discussion on testing. Thus, as a matter of self-protection, the project manager must either repeat the discussion on testing in all the applicable sections or make clear cross-references to the section which covers the material.

Table 4.1 illustrates an evaluation grid form often used by reviewing engineers to evaluate technical proposals. The grid form summarizes the evaluation factors described above for each of the major areas of the RLM simulator project. Upon completion of the technical evaluation of each proposal, the customer's project leader will total the evaluation points to obtain the relative technical standings. The table demonstrates how arbitrary technical scores of 88 for Creative Electronics Corporation, 74 for ABC Company, and 62 for XYZ Corporation might be evolved upon completion of the proposal evaluations.

The scores that were derived from the proposal technical evaluation reflect the relative excellence of each of the proposals as far as addressing the areas of the TPR are concerned. The numerical values of any low score cannot in themselves be used to render the particular proposal unacceptable. The low score is merely an indicator to identify proposal areas that require more detailed review and to ultimately establish a basis for the decision as to whether the system is

TABLE 4.1   Technical Evaluation Grid, Radar Landmass Simulator

Areas of evaluation	Weight	Scores		
		XYZ Corp.	Creative Electronics Corp.	ABC Co.
Design approach				
Function and signal flow	5	2	4	3
Computer system	5	2	5	4
Total	10	4	9	7
Real time				
Computation speed	8	5	8	6
Mathematical model	8	5	7	6
Map storage transfer	10	5	10	7
Total	26	15	25	19
Resolution				
Computer program	8	4	6	5
Computation technique	6	3	5	5
Block diagrams	4	3	3	3
Total	18	10	14	13
Radar system				
Radar controller	8	6	7	7
Simulator characteristics	8	7	7	6
Total	16	13	14	13
Flexibility				
Design features	10	7	9	8
Revision techniques	8	5	7	6
Total	18	12	16	14
Test program				
Hardware	2	2	2	2
Software	6	4	5	4
Integration	4	2	3	2
Total	12	8	10	8
Total technical score	100	62	88	74

capable of meeting the procurement specification requirements. In making determinations as to whether a proposed design is capable of meeting the performance requirements of the specification, government procurement policies require activities to establish whether a marginal or unacceptable design that is offered is susceptible to being made acceptable by reasonable revisions. Except for the most deficient proposals that indicate that the offeror does not possess basic knowledge of the required technology, meaningful discussions or technical negotiations with offerors submitting marginal or even unacceptable proposals can result in clarifying misunderstood requirements and the rendering of revisions that result in making a proposal acceptable.

4.6   Material and Hourly Effort Breakdown

A cost breakdown of each deliverable item of the procurement is generally required for negotiated contracts. Table 4.2 illustrates the type of form used for the breakdown presentation of the deliverable items. The technique for deriving estimates of hours, material, and other cost factors is discussed in Chapter 5.

The design, manufacture, programming, integration, and test of the different subsystems of modern complex equipment involve the effort of different types of personnel. The results derived from the systems engineering effort represent the basic information regarding the identity and function of the various subsystems.

In the line organization, the project manager assigns to cognizant team leaders the task of estimating the different types of man-hours, materials, and other cost factors relating to the various subsystems. The estimates are presented in a standard form to facilitate the accumulation of data for the various subsystems. The program manager is responsible for challenging estimates, eliminating redundancies, and organizing and assembling the figures into the form required by the TPR.

In the matrix, the project manager and the functional manager have the responsibility for accumulating the estimates for the various subsystems. Under their dual supervision the discipline supervisors, who are usually charged with

TABLE 4.2   Cost Breakdown of Items, RLM Simulator Project

	Item 1		Item 2	
Cost Areas	Hours	Dollars	Hours	Dollars
Material		XXX		XXX
Handling, %		XXX		XXX
Subcontracts		XXX		XXX
Overhead, %		XXX		XXX
Engineering	XXX	XXX	XXX	XXX
Overhead, %		XXX		XXX
Program design and writing	XXX	XXX	XXX	XXX
Overhead, %		XXX		XXX
Manufacturing	XXX	XXX		
Overhead, %		XXX		
Testing	XXX	XXX		
Overhead, %		XXX		
Other	XXX	XXX	XXX	XXX
Overhead, %		XXX		XXX
Subtotal		XXX		XXX
G&A, %		XXX		XXX
Subtotal		XXX		XXX
Profit, %		XXX		XXX
Total cost		XXX		XXX

the task of providing estimates for the different subsystems, present the information in the form of work packages. The work packages, which include scheduling and other types of information in addition to the cost estimates, are submitted to the two matrix managers for collating and editing. The final estimate is provided to the matrix executive for management approval. It should be noted that although the functional manager performs an important role in the estimating process, the project manager is responsible for making sure that accurate engineering, test, and other technical resource hours and facilities are contained in the cost estimate for the system. A matrix conflict will surface if differences of opinion between the matrix and the project manager over estimates cannot be resolved. Such issues will require resolution by the matrix executive.

Very often, the procuring agency requires a cost breakdown that is more detailed than just an accounting of the cost of each deliverable item. For example, if, for the RLM simulator, the TPR stipulates that the cost breakdown of the simulator be detailed by system and include a separate breakout of programming and testing, the information would be presented in a form similar to that shown in Table 4.2. The subsystems of the simulator which would be broken down by the various cost categories would include the different elements indicated in Figure 3.3.

4.7   Scheduling

The accurate scheduling of the time and effort required to complete a complex project, which often extends over a period of several years, requires a realistic appraisal of the complexity of the various tasks that must be performed, assurance that the necessary resources and competent personnel will be available when needed and a rationale for estimates of the time required to complete different tasks that is based on direct or related experiences.

A Gantt chart for each of the deliverable items of the contract schedule, identifying the period required for each different type of effort for each item, provides a comprehensive view of the project timetable. The structure of the Gantt chart for the RLM simulator is illustrated in Figure 4.1, which identifies the various functions required to produce each item.

Because of the importance of prompt delivery of the items included in a procurement, the offeror must be prepared and able to expend the required effort at a rate which will permit maintaining the delivery schedule. Quite often, this requires hiring additional personnel, procuring required capital equipment, subcontracting, or undertaking other special steps. The plans of action that would be required should be sufficiently detailed to substantiate the way in which the procurement schedule might be met.

If an offeror is in a position to improve the delivery schedule, it should emphasize this fact in its proposal and document the basis for the improved

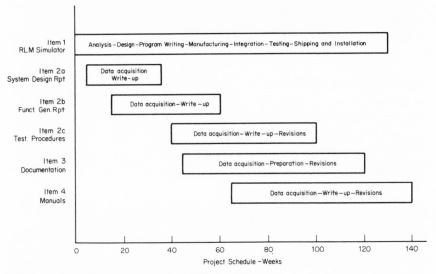

FIG. 4.1   Gantt chart structure for RLM simulator.

delivery schedule. Invariably, the company that can offer an improved delivery schedule has the necessary know-how and experience for the procurement, and its technical and cost proposal should reflect its advantages.

4.8   Project Resources

Of particular importance is the availability of the offeror's facilities and personnel for meeting the schedule of a program. One frequent cause of difficulties in meeting contract schedules is lack of availability of personnel or facilities due to the demands of other projects. The reviewing procuring officials will be interested in determining the overall schedule of the offeror's projects and its ability to handle the procurement under negotiation.

Figure 4.2 graphically portrays the engineering capacity of the Creative Electronics Corporation. The figures for engineering hours are the totals of electronic, mechanical, and other engineering disciplines. The solid line represents the engineering loading that is and will be required in order to perform under existing contracts. The dotted line represents the additional number of engineers that would be required if the company were to receive the award for the contract on January 1.

When estimates of projected effort are made on the basis of a weekly interval, the number of hours for a person is given as 40 (40 hours per workweek). There are numerous ways to show the information; in Figure 4.2 the numbers on the left side of the chart would be multiplied by 40 to indicate the number of engineering man-hours available for any particular month.

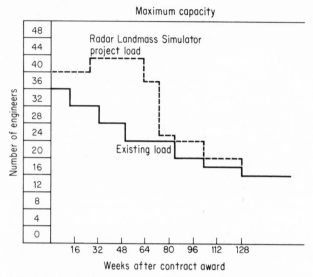

FIG. 4.2   Schedule of Creative Electronics Corporation engineering load.

Similar types of graphs should be provided to show the company's available capacity in terms of such resources as programmers and manufacturing facilities.

The information shown in Figure 4.2 serves as a very useful tool in permitting the reviewing engineer to see at a glance the ability of a company to handle a particular project as far as personnel availability is concerned. The text of the proposal should expand on the company loading to demonstrate that personnel requiring special engineering and programming experience are available to properly execute the requirements of the project under consideration.

It should be noted that in the line organization, the project manager is responsible for obtaining and collating the type of information upon which Figure 4.2 is based. In the matrix organization, the primary responsibility for providing the above-noted plant and project loading data rests with the functional manager. The project manager would have to negotiate with the functional manager and, if necessary, revise the planning schedule to develop resources that will be available in a time frame that will make possible the meeting of the various line items of the contract.

4.9   Description of Facilities and Experience

The description of the facilities should include a schematic layout of the overall plant floor space, a list of machine tools and other capital equipment, and a description of laboratory facilities and other physical property of the company. In particular, the proposal should describe in detail those facilities which are necessary to the project and which will be used to complete the project under consideration. The general theme that has been expressed previously may be reemphasized: offerors should stress any point which will contribute to placing themselves in a more favorable competitive position.

Because of the importance of the programming required for the computer approach that Creative Electronics is offering, the proposal should emphasize the credentials of the computer science and programming personnel who will be on the project team and should identify the software tools, such as compilers and analyzers, that will be utilized to develop and test the programs for the system.

The portion of the proposal relating the experience of Creative Electronics should include the following.

First, the experience of the company in completing contracts involving the same type of equipment should be mentioned. This portion of the proposal should list each such contract and briefly describe how the effort and product of each contract were or are similar to those of the desired contract.

Second, the experience of key personnel who will be assigned to the project should be mentioned. This portion of the proposal should include a brief description of what each individual's function will be in the program, each person's title, and a description of the background, experience, and qualifications of each individual which are necessary to carrying out his or her functions efficiently.

4.10 Summary

The TPR is one of several documents comprising the RFP that is distributed to qualified companies in a procurement solicitation. A TPR for a major procurement generally requires each offeror to submit a proposal in separate volumes, including volumes on engineering presentation, implementation plan, and logistics support.

The engineering presentation volume of a technical proposal is the prime document which determines whether the design an offeror proposes is acceptable or unacceptable.

Generally asked to be included in the engineering approach volume are discussions of technical areas that are considered to be critical to the equipment function or of areas that represent difficult engineering problems.

The procuring agency or organization must evaluate each technical proposal against a predetermined set of criteria to establish which company offers the best technical design for the equipment under procurement. The project manager's sensitivity to the customer's evaluation criteria makes it possible to structure the technical proposal in such a way as to satisfy the criteria, thereby enhancing the chances for a contract award.

The TPR sets forth the organization of and the type of material that is to be presented in the offeror's technical proposal. The information required usually includes the technical description, proposed schedule, cost breakdown, experience, and description of facilities. The technical description should be written in such a way as to cover thoroughly the evaluation factors that will be used, the identity of which the offeror might learn or deduce from previous similar procurements. In addition to considering the evaluation factors, the project manager must be certain that the technical description covers all the significant specification requirements as well as the points in the TPR document.

The cost breakdown involves separating the basic cost elements of material, engineering, programming, and manufacturing, which constitute the bid price. The degree of detail of the cost breakdown is specified in the TPR document, and it is important that the project manager understand the requirements and comply with them.

The cost breakdown should be presented in a chart form or on forms provided by the customer. The schedule of required effort should be presented in a Gantt chart and should clearly show the sequence of effort that is proposed for the project. In addition, the overall schedule of the total business that the offeror has, including the additional effort that the proposed project would entail, should be presented. This information should be shown in relation to the total maximum capacity that the offeror possesses in personnel and facilities.

The description of the offeror's facilities and experience should emphasize those facets and disciplines that are directly related to the proposed procurement and should show how these company resources will be applied to the proposed project.

Briefly stated, the technical proposal should cover all stipulated and anticipated areas on which the proposals will be evaluated, and the project managers must emphasize every point that will contribute to fostering a competitive advantage for their company.

The procuring activity identifies in the TPR document the information to be presented in proposals. The proposals are evaluated to establish the most highly qualified offeror. The evaluation factors established by the customer are used as criteria for the scoring of proposals and are often made known to the offerors. The project manager preparing a technical proposal should prepare an outline and make sure that all points of the TPR are fully addressed.

The proposal usually requires, in addition to the technical discussion, a detailed breakdown of material and labor costs and overhead rates. The schedule of effort proposed for meeting the delivery requirements of the equipment is also required in each proposal. To facilitate the evaluators' reading and understanding of the proposal, the material should be organized logically and written in the most lucid manner possible. Special attention must be given to the TPR to make sure that all the required information is provided in the form and sequence specified.

BIBLIOGRAPHY

Defense Contract Management for Technical Personnel, Naval Materiel Command, Washington, D.C., 1980.

5

ESTIMATING
OF COSTS

5.1 Introduction

Most companies existing in today's dynamic economy derive significant portions of, and in many cases all, their business from competitive procurements based on bidding. Thus the growth, welfare, and frequently the very existence of an organization depend on its success in accurately estimating the cost to perform satisfactorily under different contracts.

For advertised procurements in which the required items are completely defined and no development effort is involved, the bid price, which is based on the offeror's cost estimate, is usually the sole determinant of whether or not a contract award is received. For negotiated procurements, which generally involve development effort, the proposal price may be revised during the negotiation process. The project manager must know the cost threshold which represents the difference between a profitable and an unprofitable contract. An accurate cost analysis of the proposed procurement is the only means by which the necessary cost information can be established.

The importance of accurate estimates is particularly evident in procurements which involve engineering design and which are based on some form of fixed-price contract arrangement. A new design involves a degree of uncertainty in terms of man-hours or other expenditures that might be required. Adequate provisions should be made in the cost estimate to cover those areas of risk. A company which does not provide for the risks which invariably occur in new designs may be successful in obtaining contract awards because of its low bid, but it will generally experience losses on its contracts. Obviously, no industrial organization can survive long when operating at a loss. On the other hand, a company that is too conservative in its cost estimate will not receive contract awards and will wither on the industrial vine.

In estimating the cost to be quoted on a procurement, project managers must use all their experience, business acumen, and knowledge to derive an estimate which will not only enable them to get a contract award but will result in a reasonable profit for their company.

The difficulty in deriving accurate cost estimates is directly proportional to the amount of engineering design or development work that must be performed on a procurement. The first all-important requirement is that the project manager know what is required in a procurement and what design approach will be taken to fulfill the requirement.

5.2 Cost-estimating Concepts

The two basic categories of costs that must be considered are "recurring" and "nonrecurring." Examples of a nonrecurring cost would be design engineering, programming, and breadboard testing. Such costs generally are absorbed by the prototype units of the contracted equipment. A typical recurring cost would be the manufacturing cost for a system. A constant recurring cost is inherent in every unit of the equipment being produced.

Some of the factors that make up the cost estimate for a production run of an article are cost of materials, manufacturing and assembly costs, quality-assurance and testing procedures, special tooling, and overheads. In addition, the cost estimates must be modified to reflect anticipated rejection rates, learning curve impact, and other contingencies.

The cost estimate for equipment being produced from already completed engineering and development effort is basically a statistical exercise using current figures for material and labor rates and deriving the scope of manufacturing, assembly, and other types of effort from historical data on the same or similar equipment. A company that can implement some labor-saving technique or money-saving manufacturing procedure would enjoy an edge over competitors for a production contract.

The techniques and thinking process required to estimate the cost of designing, developing, programming, building, and testing a prototype device or units of unique articles differ greatly from the disciplines previously mentioned for estimating production runs. Before the project manager can begin to converge on a cost estimate, the device that is to be cost-estimated must be defined in detail. The initial steps would require answers to the following type of questions relating to the equipment in question:

1. What is it? (Derived from the specification)
2. How will it work? (Derived from the engineering analysis)
3. What are its subsystems? (Derived from the work breakdown structure)
4. What is the technical description of each subsystem? (Derived from each of the elements of the block diagram)
5. What is the scope of programming required? (Derived from computer program analysis performed for computerized systems)

Once project managers have clearly established the answers to the above questions, they can draw upon their judgment to coordinate the estimates of engineering and development hours, manufacturing hours, programming hours, and material costs for each subsystem that are rendered by various assigned personnel. Prior to applying hourly rates, overhead, and other necessary cost factors, the estimate of hours is reviewed to eliminate such factors as redundant cost elements.

5.3 Pitfalls of Estimating

The two major pitfalls of accurate estimating of costs for prototype equipment are errors in applying the mechanics of estimating and judgment errors.

In deriving the cost estimate, the project manager should review the work to be sure that none of the following errors have been made.

1. *Omissions:*   Was any significant cost element forgotten? For instance, are there any bench checks planned and does the estimate include the engineering, material, and other costs for such effort?

2. *Inadequate work breakdown:*   Does the work breakdown structure being used adequately account for all the subsystems and efforts required of the device?

3. *Misinterpretation of the equipment data or function:*   Is the interpretation of the complexity of the device accurate? Misinterpretations will result in estimates that are either too high or too low.

4. *Use of wrong estimating techniques:*   The correct estimating techniques must be applied to the device in question. For instance, the use of cost statistics derived from production runs of a similar subsystem for a prototype device which requires engineering and/or development work will invariably lead to excessively low estimates.

5. *Failure to identify and concentrate on major cost elements:*   It has been satistically established that for any piece of equipment, 20 percent of the subsystems will account for 80 percent of the total cost, as depicted in Figure 5.1. The point is that project estimators should concentrate their time and effort on

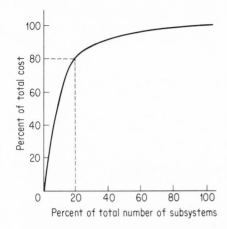

FIG.   5.1   Pareto's   law   of distribution.

the high-cost subsystems and categories of effort so as to enhance their chances for establishing an accurate cost estimate.

6. *Failure to assess and provide for risks:*   By its nature, a prototype device involves engineering and design effort as well as breadboarding and bench-testing for verification. In addition, computer programming requires a rigorous, detailed testing and verification effort during its creation in order to ensure acceptable quality. Such tests usually involve a significant expenditure of effort in redesigning and refining.

The project manager of the estimating personnel must recognize the necessity of planning for the additional effort and of using judgment in estimating the number of man-hours and the additional material that might be required to avoid underestimating the cost of the equipment.

5.4   Cost-estimating Plan

In both line and matrix organizations the elements that comprise the appropriate tier of the work breakdown structure are used as the basis for establishing the different estimating area or subsystem assignments. The line organization project manager generally has the sole responsibility of making estimating assignments, directing the effort, editing the different inputs, and preparing the collated estimate for review and approval by company executives. In the matrix, the estimation of different areas is under the joint supervision of the functional and project managers. The inputs in each area are generally provided as work packages, of which the cost estimate is one portion. The work packages provide not only the desired estimates but also availability of resources, scheduling, and other types of information that tend to support the accuracy of the estimate.

In deriving the estimate for the RLM simulator, the work breakdown structure shown in Figure 3.4 could be used to discuss the estimating plan. To pro-

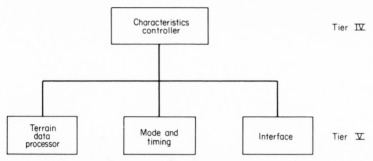

FIG. 5.2 Work breakdown structure of characteristics controller.

vide a detailed cost estimate for the characteristics controller, the system would be broken down to the next tier level in the structure. Figure 5.2 represents a detailing of one of the areas of the RLM simulation that would be analyzed for cost estimating. Further breakdown to a higher tier may be required to derive a cost estimate for creating the subsystem which will be satisfactory for bidding purposes.

Individuals assigned to estimate the cost of a particular subsystem will find that communication must be established with others who are assigned to provide estimates on the cost of interfacing subsystems in order to avoid redundancies and omissions. In addition, for a computer system, it is essential that coordination be established with the computer scientists and programmers charged with making estimates of the cost of the computer program that will drive and control the particular assigned system.

In addition to estimating the costs directly related to materials, engineering, programming, and manufacturing, other peripheral costs (such as those of data acquisition, special tools and facilities, and testing) that are related to the hardware and software portion of a particular subsystem must be identified and included in the cost estimates.

5.5 Classification of Effort

In deriving the estimate for any subsystem, it is important that the responsible individual establish whether the modules or program routine requires creative developmental effort or exists in the company library as data reflecting a previously completed development effort. It is also important to establish whether an existing design can be adapted to the subsystem under review. Development, design, and programming effort entails the expenditure of costly man-hours, and the main object of reviewing past developments is to avoid the expense of "reinventing the wheel."

In addition to the company data, it is important that outside available data sources be explored for design data so as to minimize the estimated cost and submit the most competitive bid price. (The U.S. Government maintains the Defense Documentation Center, which serves as a vast fountain of information for parties interested in any subject and possessing the required security clearance and need to know.)

To illustrate the classification of engineering effort, the characteristics controller subsystem noted in Figure 5.2 as Tier IV or the work breakdown structure is further fractured into the elements of Tier V and the type and scope of effort identified in Table 5.1.

For the elements which require development effort, a relatively large amount of engineering time would be required to complete the design. In addition, the amount of testing time would have to be high to take care of breadboarding, etc. Manufacturing effort would also be required to translate the engineering design into hardware.

For those elements already developed or available as off-the-shelf designs designated as B and C a minimum of engineering effort would be involved. Since the items would probably be procured as completed modules, the cost of the assembled unit would be much higher than that of the material that had not undergone the manufacturing and assembly effort.

Establishing the cost estimate for the computer programs is somewhat more tricky because of the nature of the art or craft of programming. Different programmers may program a computer to perform a particular function in different ways, and each approach requires a different number of man-hours. In addition, an already developed program does not possess the type of clear identity that hardware possesses, so that off-the-shelf programs that could be used for similar applications would not be readily recognized. In effect, almost all programs for new applications require a design effort. However, an organization that has accumulated a library of programs may be able to adapt one to serve a particular similar application. The estimate of the effort required to create the software for a particular module could be made with the format shown in Table 5.1, except that the categories of hourly effort would include analysis, software design, program writing, testing, and documentation. As noted for the hardware breakout (Table 5.1), the number of estimated hours

TABLE 5.1   Classification and Estimate for Hardware Effort for the Characteristics Controller

Element	Classification of effort	Engineering, hours	Manufacturing, hours	Testing, hours	Material cost, dollars
Terrain data processor	A	550	500	600	4000
Mode and timing	C	200	350	150	1500
Interface	B	350	450	300	2500
Total		1100	1300	1050	8000

A = development; B = standard engineering; C = existing design.

would be contingent on whether the software module had to be developed as a new module or an existing design could be adapted.

Since the number of man-hours that may be required to create a particular program varies among individuals, and because of the dynamic technological changes that characterize the computer hardware and software field, meaningful statistical data on which to base programming estimates are not available. At present, the number of lines of code that are required for programs is generally used as a base for cost estimates. An average of one line of code per hour is generally used, with a range from two lines per hour for the proficient programmer to two hours for a single line for the slower programmer.

Other factors that should be considered in establishing an estimate for creating a program are the quality of available resources and the number of program instructions that are required.

Special attention is required for estimating the number of man-hours required for testing and documentation of computer programs, since most cost overruns for computer systems have been due to underestimates of those two types of software effort.

5.6 Management Review of Engineering and Programming Effort

In the line organization, the project manager is responsible for the collation and review of estimates. Revisions are made primarily to eliminate duplication and to make sure that the pitfalls discussed in Section 5.3 are avoided. The manager also challenges any estimate which appears to be too high or too low.

The functions of the project manager in the matrix are essentially the same as those of the project manager in the line organization, except that the review effort of the former is implemented as a joint effort with the functional manager.

When the estimates of engineering and programming hours for various subsystems or elements are submitted by the different engineers, the review by the group leaders, project and functional managers, and higher management officials includes the following.

1. Review all systems to identify identical elements for which redundant engineering charges are estimated.

2. Review all systems to identify elements for which a design may have been accomplished on other projects, thereby making available an off-the-shelf design instead of expending a duplicating engineering effort on the current project.

3. Review all systems to identify elements which, although different, may be sufficiently similar to warrant adopting one standard element for a maximum number of systems without seriously compromising the performance characteristics of the hardware.

4. Review all systems to identify elements which may be sufficiently similar

to elements of an off-the-shelf design to make adoption of such an off-the-shelf design possible without the performance characteristics being compromised in any significant way.

5. Review available sources of information to determine whether any design data relating to the project on hand is available for use by the company.

5.7 Material Costs

The estimate of material costs is a fairly straightforward effort, normally based on the current level of market prices. For projects spanning a long period, the material costs may be modified to anticipate the price that the company expects to pay at some future date. The modification factor is usually based on price indices and trends. However, for normal short-term projects (1 year or less) the current market prices are used. These costs must include factors to account for waste, spoilage, and general fabrication errors normally referred to as "tares." Thus lead engineers should in their estimate of material costs add a quantity to account for tares, based on a statistical curve derived from past experience with a particular component.

The curve reflecting the spoilage of any particular element assumes the general shape of an exponential curve illustrated in Figure 5.3. The curves will assume different shapes for different elements and different histories of experience. For instance, curve *a* in Figure 5.3 might be for a type of transistor not generally used by Creative Electronics. Therefore the spoilage rate on the basis of 200 units would be 20 percent. For 1,000 units, the spoilage rate would be 100 units, or 10 percent spoilage.

The spoilage curve for a standard electronics component tube used more or less on a regular basis by Creative Electronics might follow curve *b* on Figure 5.3. The rate on the basis of using 200 units would be 10 percent spoilage, and for 1,000 units the rate would be about 5 percent.

The slope decrease or flattening out of any curve as the quantity of units used is increased is due in part to the experience or learning factor that is realized,

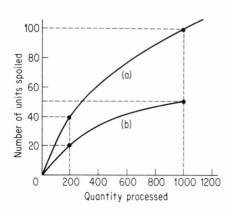

FIG. 5.3   Curves of spoilage vs. quantity produced.

thereby cutting down the spoilage rate for larger quantities. This is a reflection of the philosophy that if someone makes a mistake resulting in spoilage of an element due to carelessness or other shortcomings, that person will make a special effort not to repeat the mistake. Theoretically speaking, if a person performs a certain task with certain materials often enough, a point will be reached where no mistakes are made or where the spoilage curve is horizontal.

Further, the shop supervisor and production manager will use the mistake of one worker which caused spoilage to teach other workers handling the same element to avoid duplicating the unfortunate experience. Other steps, such as providing special jigs, fixtures, tools, and procedures, will be instituted to provide for a more efficient and less wasteful operation.

5.8 Estimating Manufacturing Costs

When the estimated engineering effort and material costs are gathered and correlated by project managers, their next step is to determine the estimated manufacturing costs for the project. The production manager is the key individual for estimating manufacturing costs, and the detailed procedures for gathering such costs vary for different organizations.

Essentially, the procedures for getting estimated manufacturing costs involve the following basic functions.

1. Providing the production manager with information relating to the hardware to be produced
2. The establishment by the production manager of the basic areas of effort
3. The solicitation by the production manager of the cost estimates of different production supervisors
4. Estimation of effort by the different production supervisors
5. Accumulation and modification of production costs by the production manager
6. The transmission of cost estimates to the project manager

The information required by the production manager includes a comprehensive description of the physical design of the different systems, their relation to each other, and the overall hardware. In addition, a description of the specification requirements which relate to the hardware must be provided and interpreted for the use of the production manager. This is particularly important if the equipment must be subjected to environmental tests, such as vibration shock, since special manufacturing process would then be involved.

The production manager categorizes all the types of production effort and disseminates the information to the various shop supervisors for detailed production estimates. Examples of categories of manufacturing effort are assembly wiring, sheet metal work, and cabling.

The shop supervisor of each production area estimates, on the basis of the descriptions given by the production manager, the production hours required for this area of responsibility and transmits these figures back to the production manager who, in turn, correlates all the figures and derives an estimate for the number of production hours required to complete the project.

The cycle by which the production hours are derived is completed when the production manager provides the project manager with these figures.

5.9 Estimates of Schedule Side Items

The primary procurement item, which is the equipment, dominates the interest and planning of the individuals concerned with a project, often at the expense of the supporting or side items listed on the schedule. In the case of the RLM simulator, cost estimates must be derived for the engineering reports, engineering and programming documentation, and installation and maintenance manuals. Other types of support items that are often required are spares, maintenance services, special tools, engineering drawings, etc. The cost of support items represents a significant part of the total procurement estimate. Cursory consideration can result in an underestimate which could lead to an unprofitable contract.

The engineering reports are usually created by the company individuals who are involved with the design and engineering of the different systems that make up the equipment. Methods of coordination of the various inputs vary from one organization to another, but it is the project manager who must make sure that the content and format comply with the specification requirements. Estimates for the preparation of engineering reports are rendered by the group leaders for different subsystems and are collated by the project managers.

The engineering drawings are usually created by the drafting personnel, who work closely with the design engineers. Estimates are based on the complexity and size of each drawing, which determine the man-hours required for the effort.

Documentation relates primarily to the programming effort and must provide data of sufficient detail to enable reviewers, users, and other concerned personnel to understand how the program is designed and to make desired revisions. The data in the documentation category includes such items as program descriptions, program performance, specification, software engineering notes, program design language, coding standards, and test procedures. Estimates for the documentation that the contract requires are compiled by the chief programmer and submitted to the program manager to be included as an element in the proposal bid.

Installation and maintenance manuals usually take the form of texts. Publishing firms that specialize in the manuals for complex systems are generally subcontractors to the prime contractors. Offerors will solicit estimates for manuals from the publishing firms; these estimates are based on descriptions of the technical content, the manuals' specifications, the detail required, the number and type of illustrations, etc. During the pre- and postcontract phases, a close liaison between the two organizations is essential to make sure that the manuals are acceptable to the procurement agency.

5.10   Application of Overhead Rates

The different overhead rates and G&A (general and administrative) rates are applied to the direct costs that have been gathered and modified to evolve an estimate for the total project cost.

In Table 4.2 is shown a typical organization of the major cost elements of two items of the RLM simulator contract. The total cost would assume the same form but would summarize the individual items. The overhead and G&A rates are a function of the past experience of the company and vary significantly among companies, geographic areas, and industries. The rates are adjusted periodically to reflect the latest company experience.

It should be noted that a company operating at full capacity will experience minimum overhead rates, whereas a company which has a significant amount of excess capacity will experience high overhead rates. The reason for this is that overhead burden is essentially fixed and that the costs (such as heating, lighting, and real estate taxes) are prorated over the in-house business. There have been cases of companies having essentially one job in-house that consumed a fraction of the available capacity so that the entire overhead burden was loaded on the one job. In such cases, the overhead rates are extremely high. Basically, the subject of the establishment of rates for overhead, G&A, etc. is a matter for accountants and auditors. Project managers need only to be aware of their existence and their effect on the project. However, there is very little that they can do to influence them.

5.11   Commercial Considerations

Theoretically speaking, the estimated project cost (including profit) is identical to the target bid price for a project. However, such an ideal situation rarely occurs, since in the free competitive market the commercial considerations must be recognized even when bidding on a government procurement opens only to a limited number of qualified companies.

The final revision of the derived project costs is dictated by such commercial considerations and is made by the highest management echelons of a company. These considerations involve too many facets to discuss them separately in detail. However, the general categories of consideration that come into plan are covered in the following breakdown: type of effort required, type of contract solicited, type of competition, and the company's need for additional business.

The degree to which management is willing to modify the estimated project cost downward is a function of the risk of sustaining a loss on the contract that the company is willing to assume by submitting a low bid. On the other hand, the management of a company may, for different reasons, feel that it enjoys a position of advantage in a procurement and therefore will be willing to assume the risk of not getting a contract award as a result of raising the estimated project cost in order to realize a greater profit. This basic philosophy, however, is contaminated by cross-currents of other factors, such as incentive and penalty clauses and profit sharing. The basic principle should, however, be kept in mind.

The risk factor associated with different types of contracts is discussed in Chapter 7. To cite the two extremes, a company on a cost-plus-fixed-fee procurement can afford to reduce its cost figures by a significant amount without assuming any risk or loss, since all the incurred costs are reimbursable. On the other hand, if the procurement is a straight fixed-price contract, the maximum risk would be incurred if a reduction in cost were made.

The nature of competition and type of contract help determine whether the estimated cost figure should be revised. If Creative Electronics has reason to believe that a competitor with a history of submitting low prices on procurements is one of the companies solicited and if the procurement is considered to be worth the risk, management may shade costs to a point which is thought to be competitive with the anticipated price of the competing company. The company's existing plant loading, including engineering work, will be a determinant as to how much risk the company is willing to assume in order to obtain additional business. Basically, a variation of the forces of supply and demand enters the picture. If there is excess capacity in the plant, a company should be willing to assume more risk by adjusting its bid downward than it would if all the facilities of the company were being operated at full capacity.

In an analysis of how much of an adjustment is to be made in the estimated cost figures to arrive at a bid price, consideration must be given not only to the present commercial status of the company but to the status projected for the future. For instance, the RLM simulator program is scheduled as a 130-week effort, and the workload demands on the engineering and technical resources of Creative Electronics that would have to be handled by the contract are illustrated in Figure 4.2. In the case of manufacturing, the fabrication shop may be loaded for the next 50 weeks, but after that, the projection indicates a decreasing load. It would be essential to the company to obtain the contract award for the RLM simulator in order to place work in the shop 50 weeks from the anticipated date of contract award.

The same factors just discussed would be considered in establishing the ceiling and incentive arrangement to be included in a bid proposal. The actual mechanics of the fixed-price incentive contract (FPI) are discussed in Chapter 7. Any adjustments in the cost figures instituted by management must be reflected as adjustments in hours of effort or material costs in order to provide a consistent picture. It should be recognized that the adjustments of cost estimates by management when deemed necessary to derive a bid price are very legitimate and necessary as far as the existence of the company is concerned. A downward adjustment in estimated cost would not be executed in a completely arbitrary mannner but would be derived after consultation with the people in charge of the affected areas of effort. Usually, it is found that a limitation imposed on one's area of effort stimulates one's imagination and resourcefulness to seek cost-cutting steps in order to operate within set budget figures.

5.12 Summary

Accurate estimating is essential for a company's success in negotiated procurements involving development effort even though the proposal bid price is subject to revision during the negotiation process. The project manager's knowledge of the project's estimated cost determines whether the procurement represents a profitable or a losing effort.

The project manager must have a comprehensive knowledge of the design, the scope of different type of effort, and other factors related to cost in order to carry out accurate cost estimating. The most common pitfalls leading to erroneous estimates have been presented in this chapter. Statistically, it has been established that for development procurements, approximately 20 percent of the subsystems represent 80 percent of the cost of a system.

The work breakdown structure is a common tool for assigning the cost-estimating areas to different groups. The areas are then translated into block diagrams down to the tier level required by the estimator. To the degree possible, subsystems that have already been designed should be used or adopted to avoid the unnecessary duplicating of engineering effort. To facilitate the identification of subsystems requiring development effort or of subsystems that can be adopted, the different element areas should be analyzed and classified and estimates should be established for man-hours and costs.

The estimate for the creation of computer programs requires special analysis because of the types of effort that programming entails.

The evaluation of risk for different subsystems that are to be developed must be established and reflected in the estimates. The project manager's accurate judgment as to the magnitude of the risk factor is an essential ingredient in formulating realistic estimates for developmental efforts.

Discussion of the estimates that are required for support items, material costs, and the application of overhead rates has been presented to complete the cost-estimating discussion.

BIBLIOGRAPHY

Farr, L., and B. Nanus, *Factors that Affect the Cost of Computer Programming.* Report EDS-TR-64-448, USAF Electronic Systems Command, Washington, D. C., 1964.

6

CONTRACT
SCHEDULE

6.1 Description

An RFP on any procurement will include a proposed contract schedule which forms the basis for the final contract. A contract schedule is generally divided into sections, each setting forth the specific language and terms which establish the contract requirements.

The words "contract schedule" might mislead an individual into concluding that the document applies only to times of delivery. The references to the calendar are only one part of the contract schedule.

Far more significant from the contract point of view, and therefore to project managers and their companies, are the contract clauses, terms, and other conditions which are in the schedule. Usually any issue of dispute between the two parties in a contract revolves around the application or interpretation of some element or phrase of the schedule language.

The contract schedule is the most important legal document of a procurement and takes precedence over any other document in the event of any conflict of language. Usually, the order or precedence of documents relating to a procurement is cited in one of the sections of the schedule in order to preclude any misunderstanding in conflicting areas.

The following is a condensed illustration of the different sections of a contract schedule that would be applicable to the RLM simulator project.

SECTION A

Item	Contract Items	Quantity
1	Radar Landmass Simulator, Device 5A1 delivered and installed at site	1 each
2	Engineering reports (government allowed 30 days for each review and comments; approval to be rendered when acceptable)	
	(*a*) System design	2 copies
	(*b*) Data storage and processing	2 copies
	(*c*) Testing procedures	2 copies
3	Technical/administrative monitoring documents	2 sets
4	Support documentation listed in Appendix B	2 sets

SECTION B   DELIVERY

The following items are to be furnished under Section A above in accordance with the following schedule at the destinations noted below:

ITEM 1: One unit shall be delivered and installed by the contractor and made ready for acceptance 195 weeks after date of contract at the Naval Training Center, Orlando, Florida, Building 7852. The government will take five (5) weeks for performing acceptance tests. The contractor is responsible for correcting any deficiencies identified during acceptance testing and is responsible for any delays in trainer acceptance due to the deficiencies.

The following items shall be delivered to the Contracting Officer, Naval Air Systems Command, in the quantities and during the periods listed below:

ITEM 2a: Shall be delivered 35 weeks after date of contract.
ITEM 2b: Shall be delivered 60 weeks after date of contract.
ITEM 2c: Shall be delivered 100 weeks after date of contract.
ITEM 3:   Shall be delivered 120 weeks after date of contract.
ITEM 4:   Shall be delivered 140 weeks after date of contract.

SECTION C   DESCRIPTION OF ITEMS

Item 1 shall be in accordance with Performance Specification 1001, Radar Landmass Simulator, Device 5A1, and all amendments thereto.

Item 2 shall be in accordance with U.S. Naval Training Command Specification MIL-5071, Standards for Engineering Design Reports.

Item 3 shall be in accordance with U.S. Naval Training Command Specification MIL-904, Programming Manuals.

Item 4 shall be in accordance with U.S. Naval Training Command Specification MIL-1704, Requirements for Installation and Maintenance Equipment Manuals.

SECTION D   GOVERNMENT-FURNISHED PROPERTY

The following items will be furnished to the contractor by the government in the quantities and at the times indicated:

Item	Description	Quantity	Delivery to contractor from date of contract
1	AN/APQ-28 Radar Display Unit Part APQ-28-64	2	40 weeks
2	AN/APQ-28 Radar Control Box	1	40 weeks
3	Design Data on AN/APQ-28 Radar System	Lot	15 weeks
4	Cultural, Terrain-Elevation Map Details of area to be simulated	Lot	10 weeks

SECTION E   DESIGN APPROVAL

The formal approval of the Engineering Design Reports required in item 2 shall constitute authorization for the contractor to proceed with the detail design, manufacturing, and other effort necessary to meet the requirements and schedule of the contract. The government shall approve or reject, with comments, any submission within 30 days. In cases of rejection, the contractor shall submit a revised report within 30 days after notification of rejection.

SECTION F   PRECEDENCE OF DOCUMENTS

In the event of any conflict, the following is the order of precedence of documents relating to this procurement:
(1) Contract Schedule
(2) Reference Standard Contract Clauses
(3) Performance Specification No. 1001 for Device 5A1
(4) Statement of Work
(5) Referenced Specifications of Performance Specification
(6) Contractor's Proposal

SECTION G   INSPECTION AND ACCEPTANCE

Item 1 shall be set up and dynamically activated at the contractor's plant for preliminary inspection prior to shipment to the site destination. Preliminary inspections shall be made by the Contracting Officer's representative to verify that the performance characteristics of the Radar Landmass Simulator are in accordance with the contract requirements and shall require no more than 3 days if no major deficiencies are observed. The government reserves the right to direct the correction of deficiencies and to conduct subsequent in-plant inspections of the unit prior to the trainer being shipped. Contractual acceptance of item 1 will be made by the Contracting Officer's representative after delivery and installation are satisfactorily completed. Acceptance tests will be made in accordance with the Acceptance Test Procedures Report. However, the government reserves the right to perform any additional tests which are considered appropriate to verify the performance characteristics of the simulator and to direct corrective action by the contractor if deemed necessary.

SECTION H   LIQUIDATED DAMAGES

In case the contractor fails to deliver item 1 on or before the date specified in Section B of this contract, the contractor shall pay to the government as liquidated damages the sum of $100 for each day of delay up to a maximum of 90 calendar days and the sum of $50 per day for the succeeding 90 days.

SECTION I SITE AVAILABILITY

The government will have an air-conditioned room 20 by 30 feet in size, 8-foot ceiling supplied with 100 kVA, 120/208, 3-phase, 4-wire, 60-hertz power available to house and operate the Radar Landmass Simulator described in item 1 at the Naval Training Center in Orlando, Florida, no later than ten (10) months after date of contract award.

6.2 Analysis of Contract Schedule

The contract schedule, in conjunction with the specification, statement of work, and referenced documents, establishes what is required, the delivery dates, the quantities, and the conditions relating to the procurement. It is essential that the project manager make a complete and detailed analysis of the various sections of the schedule, identify potential problem areas, and evaluate the impact of any contractual requirement on the company's ability to satisfy the contract and make a profit.

In analyzing the various sections of the schedule, the following questions must constantly be kept visible to all members of the project team:

1. What is required?
2. When is it required?
3. Will the company have the necessary resources and facilities to produce what is required on schedule?
4. What contractual conditions may impose special demands or problems on the company?
5. What contractual obligation does the customer assume?

Section A of the schedule lists the various "line" items of the proposed contract and the quantities required. Items 2, 3, and 4 are commonly referred to as support or side items, and the amount and type of effort required to produce and deliver such items usually is significant. There have been numerous cases when project managers, in their zeal to concentrate on the primary hardware items, gave the support items secondary attention, with the result that the amount of money allocated to produce the support items proved inadequate. Thus an otherwise profitable contract caused a loss for the company.

The delivery schedule of the contract items is contained in Section B. It should be kept in mind that the delivery requirements specified by the procuring agency or customer have been carefully thought out and are rarely unreasonable. Usually the schedule is based on actual performance by some company as concerns similar or related items. Project managers can assume with a great deal of confidence that even if they feel that a particular delivery schedule is unreasonable, there are always competitors who do not share their view.

However, since it has been previously established that Creative Electronics is qualified for this procurement, the delivery schedule can be established as being reasonable. The project manager will be required to portray the project schedule graphically during discussions with management as well as with subordinate personnel. Figure 4.1 gives a good example of the type of schedule Gantt chart that would serve the project manager's requirements.

In deriving the type of Gantt chart outlined in Figure 4.1, the project manager should divide each contract item into segments of effort. Since the delivery of each item is fixed by the terms of the contract schedule, the best approach is to start from the delivery dates and work backward. Overlaps of different time segments reflect a buildup and tapering off of the various types of effort. The different segments for item 1, the RLM simulator, are illustrated in Table 6.1.

In like manner, the other contract line items can be shown on Gantt charts with applicable segment efforts. It should be noted that aside from general planning and setting up of personnel assignments, active expenditures of effort do not take place until an appropriate time after the contract award date. Active effort on the program writing, for instance, cannot be initiated until sufficient progress has been made in the design effort to make documentation of the drawings possible.

Of particular significance is the fact that the customer (in this case, the government) is contractually obligated to deliver data and equipment by specific dates. The obligation of the government to deliver acceptable equipment and the government's responsibility for the support of such equipment are set forth in the DAR (Defense Acquisition Regulations)° and other similar documents. It is vital that the project manager know precisely the responsibilities of the government in this area. If the customer is some agency or company for which the DAR does not apply, then the responsibilities of the customer must be specifically set forth in the contract.

TABLE 6.1   Phase Schedule, RLM Simulator

Phase segment	Period of effort, weeks
Ship and install	10
Acceptance tests	15
Integration	25
Verification	30
Manufacturing	25
Program writing	40
Design	25
Analysis	20

°The Defense Acquisition Regulations (DAR) document replaced the Armed Services Procurement Regulations (ASPR) in 1978. Eventually, the Federal Acquisition Regulations (FAR) will replace the DAR.

Section C of the contract schedule, Description of Items, identifies each of the deliverable items to be procured.

Item 1, the RLM simulator, has been described and analyzed in previous chapters.

The requirements for each of the items are called out in the specifications and documents cited for each item in question. The specifications for the drawings and reports, manuals, etc., dictate format and content requirements. The project manager must be certain that the individuals or companies assigned the responsibility for the reports, manuals, and drawings are intimately familiar with the requirements that are applicable. Failure to comply with the specification details will invariable result in rejection of the items, thereby necessitating expensive and time-consuming resubmissions. There are many case histories of companies that have experienced losses on contracts because they failed to give proper attention to the detailed requirements of the side items, such as manuals and drawings.

Section D, Government-furnished Property (GFP), describes what is to be furnished by the government for use on the contract. The government has a very real responsibility to provide items that are completely checked out and operating in a fully satisfactory manner by the dates indicated. Contractors assume the responsibility for normal diligence and care in handling and maintenance of furnished equipment, but they are not responsible for major equipment failures which are beyond their control.

Thus the government's responsibilities in relation to meeting delivery dates, providing acceptable equipment, and providing adequate and timely support in the event of equipment breakdown are highly important as far as contractor performance is concerned. Many disputes involving large sums of money revolve about the following point: How valid is the contractor's claim that the government's (or customer's) failure to meet its obligations in providing government- or customer-furnished property caused the cost overrun and/or slippage in delivery? The issues in such claims are rarely clean-cut, and contractors will invariably be awarded consideration if they can prove that customers failed to meet even a part of their responsibilities.

Section E, Design Approval, concerns the approval of the various reports for different systems. It should be noted that according to Section A the customer is under obligation to review and submit comments or approval within 30 days after receipt of each report from the contractor. The dates of approval are extremely important to the contractor, since they establish when final detail designs can be completed and when drafting, production, and other similar tasks of the program can be initiated.

If for any reason the customer fails to submit its approval or comments on a report as required, the contractor has a legitimate claim for schedule slippage or cost increase. On the other hand, it is incumbent on the contractor to submit reports that comply with the requirements of the report specification cited in

Section C. Usually specifications for items such as reports are very broad in scope and are subject to interpretation. To avoid any conflicts, it is essential that the project manager maintain close liaison with the customer's project manager in order to reach a complete meeting of the minds as to what is required. If a report is disapproved because of failure to meet any of the specification requirements, the contractor must resubmit a revised report as soon as possible and seek means to make up for any lost time due to the delays in obtaining report approval.

In any contract requiring research and development, it would be impossible to list every element and facet of the items to be delivered. Because of this fact, many areas are open to interpretation. It is essential that good faith and integrity be exhibited by both the contractor and the customer. If the contractor sincerely tries to live up to the requirements of the contract, usually the customer and the customer's project manager are willing to interpret the specification requirements to the advantage of the customer if by so doing, the end product would not be compromised in any way.

Section F, Precedence of Documents, sets forth which of the contract documents would take precedence in the event of any conflict in terms of language between or among two or more documents. The precedence sequence noted is typical for a negotiated procurement that is the basis for the simulator case. One advantage of having the contract schedule carry the highest precedence is that a particular requirement feature of the proposed contract item may be changed by mutual consent during negotiations. The revision, which would differ from the specification and/or proposal, would be described in a special section of the contract schedule and would therefore supersede the specification and proposal language.

Different types of contracts might dictate variations from the precedence of documents. For instance, by definition a two-step procurement involves the incorporation of the contractor's proposal as the contract document instead of the specification. Therefore, the proposal might be designated as the contract document with the higher precedence.

The precedence of material that is established in a contract can be vital to both of the contracting parties. If, for instance, an offeror proposes a feature that is not consistent with the specification requirement in a particular area and the contract award is made without rectifying the inconsistency, the customer will be contractually bound to furnish something which was never intended. Because the contract cited the specification as having precedence over the offeror's proposal, the offeror (who is now the contractor) may have to provide a feature more expensive than what was proposed. It is important that the schedule, which is the document having the higher precedence, address any inconsistencies between the specification and the proposal submitted by the contractor so that both parties recognize exactly what is called for in the contract.

Section G, Inspection and Acceptance, establishes the requirements relating to the testing of the primary procurement item—namely, the RLM simulator. The established ground rules of the inspections and acceptance provide protection for the contractor as well as the customer. The contract section establishes the criteria for acceptability and the periods during which tests are to be performed. Since the contractor does not receive final payment until after the articles being furnished are successfully tested and judged acceptable, it is essential that all aspects of the inspection criteria and periods be fully described in the contract in order to avoid controversy and unnecessary expenditures due to misunderstandings between the customer and the contractor.

Section H, Liquidated Damages, is one type of clause that contractors prefer to avoid. A liquidated damge clause in a contract forces contractors into a position wherein their failure to satisfy the basic performance and delivery requirement would cause them to suffer specific monetary penalties. A customer may impose liquidated damages based on delivery when the failure of a contractor to meet a specific delivery date will result in the customer's suffering a hardship or loss. Project managers, prior to permitting their company to become party to a contract, must first assure themselves that precautions against all the risks and contingencies that can be foreseen are taken by their company.

Section I, Site Availability, imposes on the customer the contractual obligation to have the housing and other facilities available by a specific date to permit the contractor to continue with the work under contract. Failure of the customer to meet these responsibilities under this section would probably work a hardship and subject the contractor to an expense which would be a valid basis for a claim by the contractor.

The various types of contract sections discussed above typify the content that might be contained in a contract. Since each procurement, whether by a government agency or by a commercial firm, is unique, the content and terms of each contract would be tailored to the procurement and would be different. It is incumbent upon the project manager to analyze all sections of the proposed contract schedule to make sure that no requirement is overlooked. Failure to recognize the possible impact of a contract clause, such as liquidated damages, could very easily be the cause of severe losses on the contract.

6.3 Summary

A proposed contract schedule usually accompanies RFPs on major procurements and eventually is incorporated as a contract document. The purpose of the contract schedule is to identify precisely what is to be procured, how many, and when; in addition, the contract schedule sets forth all the terms and conditions of a procurement contract. The contract schedule is the most important legal document of a procurement and generally takes precedence over all other documents, such as the specification. Most contractual and legal disputes on a procurement revolve around the interpretation and wording of the contract schedule.

BIBLIOGRAPHY

Defense Acquisition Regulation, U.S. Department of Defense, Washington, D.C., 1978.

7

TYPES AND ANALYSIS OF CONTRACTS

7.1 Basic Procurement Concepts

The two basic methods of procurement that are used by the U.S. government and most other public and private organizations are through formal advertising and through negotiations. In addition, procurements incorporating features of both types are used with increasing frequency by the government for major high-technology systems.

Contract awards for formally advertised procurements are usually based on price alone, whereas awards for negotiated procurements are based on discussions and ultimate agreement on many factors of a project, such as price, delivery, design features, and other relevant points.

The method of procurement will dictate whether the fixed-price or the cost-reimbursable type of contract will be used. Generally, a formally advertised procurement will use a firm fixed-price type of contract, and a negotiated or hybrid advertised-negotiated type of procurement will involve a form of either the cost-reimbursable or the fixed-price-incentive type of contract.

In the fixed-price contract, the cost is established at its signing and remains fixed provided that no changes in requirements are made and the procuring activity lives up to the terms of the contract (e.g., delivery of government-furnished property as scheduled).

In the cost-reimbursable contract, the contractor is reimbursed for all or part of the costs incurred in the execution of the contract.

Within the framework of each of the two categories, many varieties of contracts exist, each of which is applicable to the circumstances related to a particular procurement. Whereas these two basic types of contract are more commonly used for government procurement, they are often also used for procurements by corporations and other commercial firms.

7.2 Determinant Factors for Types of Contracts

Many factors influence the procurement approach and the type of contract to be used for a particular procurement. Some of these factors are:

1. The complexity of design and the type of equipment that is required. For example, a highly complex development design would require the negotiation of a cost-reimbursable type of contract.

2. The amount of risk which the contractor would assume. For example, a procurement requiring research effort in new areas involves considerable risk and would normally call for negotiation of a cost-reimbursable type of contract.

3. The period of contract life. A procurement that extends over a long period would be subject to variables such as rate changes, overhead variations, and other unpredictables. A redeterminable type of fixed-price contract which would protect the contractor against increased costs and provide the customer with any credits when decreases occurred would normally be negotiated.

4. The competition for a soliciation. If a procurement which might normally be the subject of a cost-plus-fixed-fee (CPFF) type of contract interests a number of companies, the customer can negotiate a fixed-price type of contract, which is more desirable from the customer's point of view.

5. A procurement that is based on a definitive specification. When the specification fully describes the required equipment that can be provided by two or more offerors, and when the contract can be awarded, on the basis of price alone, the procurement is normally handled as an advertised solicitation based on a firm fixed-price contract.

Other factors, such as the demonstrated performance of contractors, difficulty in estimating the costs of a procurement, the urgency of the requirement, and the contractor's accounting procedures, also influence the type of contract to be adopted.

7.3 Fixed-price Contracts

The fixed-price contract reflects an agreement by which the contractor is obligated to deliver the items as described in the specification and the contract schedule for a specified price. Precluding any changes in the contract or specification requirements, the contract price will remain fixed for the life of the contract.

However, for any major procurement involving engineering and/or development, there are almost always factors which act to mar the theoretically perfect fixed-price contract and which require consideration as far as effect on contract price is concerned. Some of these factors will be discussed in more detail.

The basic fixed-price contract is referred to as a firm fixed-price contract. Whereas a firm fixed-price arrangement imposes the greatest risk, it also offers the contractor the incentive and opportunity to realize the greatest profit. In effect, the contractor shares 100 percent in any savings due to cost reductions resulting from his efforts. The firm fixed-price contract is applicable when the purchased item can be identified in detail and when price competition exists. If any technical unknowns exist, the firm fixed-price contract is not an attractive vehicle for a contractor.

The variations of the fixed-price type of contract are fixed-price with escalation, fixed-price with redetermination feature, and fixed-price incentive.

Fixed-price with Escalation Contract

The fixed-price with escalation type of contract is used to protect the contractor against increases in labor rates, material, overhead, and other costs that are involved in performing under the contract. The escalation feature is usually tied to some offical index, such as labor rates for the area and basic raw material prices. This type of contract is applicable when very long delivery schedules are involved.

Fixed-price with Redetermination Contract

The fixed-price contract with the redetermination feature is applied where large quantities of complex equipment are procured. Its use is justified in cases not only in which the engineering and design effort is difficult to estimate but also in which the production effort and materials for the large quantity of deliverable items are indeterminable. The terms of the contract can be worded to protect the buyer or the seller or both. Probably the most widely known examples of the redeterminable fixed-price contracts are those instituted during World War II, when the armed forces required that complex weapons be designed and produced in large quantities on an urgent delivery schedule. In many of those cases, the United States government conducted redetermination negotiations for years after the conclusion of hostilities and the termination of the contracts.

There are several variations of the fixed-price contract with redetermination clauses. The following summarizes some of the different conditions that can be incorporated in this type of contract.

1. *Upward and downward adjustment:* Redetermination proceedings are conducted to establish a more accurate contract cost figure based upon cost experienced after the contract has run for a significant period of time. Both the contractor and the customer are afforded protection with this arrangement.

2. *Downward adjustment only:* This type of redetermination is known as a maximum-price contract and protects the customer. The application of this arrangement occurs when the contractor has been permitted to incorporate large contingency figures in the contract price. Any redetermination negotiations that are conducted seek to identify which of the contingencies were valid.

3. *Redetermination during the life of the contract:* As a general rule, it is desirable to establish an accurate contract price as soon as is feasible. The time of redetermination proceedings can be established as of a specific date, a particular point in the contract life, or any other agreed-upon arrangement. In essence, what is desired is to establish a time in the contract life when sufficient cost figures have been accrued which can be used for establishing a contract price. An example of 2 and 3 above might be where a contractor had included a contingency to provide a 50 percent redesign effort. At the conclusion of the design, if it were found that only a 25 percent redesign was necessary, the redetermination would reduce the contract cost by an amount equal to the 25 percent unconsumed engineering redesign cost. Further, because only half of the redesign contingency was utilized, it would be logically deducted that less drafting or laboratory developement time would be required and that an anticipated reduced effort and cost in these areas could be eliminated, thereby reducing the contract.

4. *Redetermination after completion of the contract:* Since redetermination negotiations are usually involved and time-consuming, the customer or contractor may not be able to enter such negotiation during the life of the contract—e.g., in time of war. Therefore, redeterminations are stipulated for some postcontract period.

Other variations of the fixed-price with redetermination features can be implemented to fit the requirements of the procurement and the times.

Fixed-price Incentive Contract

The fixed-price incentive (FPI) is a form of contract used for major procurements which extend over a long period of time and/or when large production quantities are involved. Its aim is to reward or penalize contractors on the basis of their ability to control costs, primarily in manufacturing and general administration. The FPI contract is not applicable where a significant amount of research and development (R&D) effort is required or where the major cost element is for materials.

The FPI contract is set up in such a way that a target cost and a target profit (generally 10 percent) are established. In addition, a contract ceiling price and price adjustment formula are established during negotiations between the contractor and the customer.

The understanding of the somewhat complex FPI contract can be enhanced by setting up a hypothetical case such as the following.

Target cost	$100,000
Target profit (10%)	10,000
Ceiling cost (125%)	125,000

The cost-sharing arrangement is 80–20 above and 90–10 below the target cost. Assume that the contractor experienced an overrun of 20 percent in the contract cost. Under the terms of 80–20 for overrun costs cited above, reimbursement would be broken down as follows:

Target cost	$100,000	
Plus 80% of 20,000 (portion of overrun assumed by customer)	16,000	
Total reimbursement costs		$116,000
Target profit	$ 10,000	
Less 20% of 20,000 (portion of overrun by contractor)	4,000	
Total profit		6,000
Total reimbursement		$122,000

Percent profit $\dfrac{6,000}{116,000}$ = about 5%

If, on the other hand, the contractor completed the contract, incurring costs of $80,000 ($20,000 less than target cost) under the terms of the cost-sharing arrangement 90–10 below target cost, the following would evolve.

Target costs	$100,000	
Less 90% of 20,000 (customer's portion of saving)	18,000	
Total reimbursed costs		$82,000
Target profit	$ 10,000	
Plus 10% of 20,000 (contractor's portion of saving)	2,000	
Total profit		12,000
Total reimbursement		$94,000

Percent profit $\dfrac{12,000}{82,000}$ = about 15%

In no case will the contractor be reimbursed in an amount exceeding the 125 percent ceiling ($10,000 × 125% or $125,000). The $125,000 ceiling figure includes all costs incurred, including profit.

It should be noted that the realized profit is expressed as a percentage of the cost of the contract.

7.4   Multistep Source Selection Procurements

In order to realize the advantages of features associated with both advertised- and negotiated-type procurements, multistep contracting procedures are frequently used by government activities, when appropriate. The two most common multistep procurement selection types are the two-step and the four-step. These types of contractor selection procedures are generally used for projects that do not involve significant research or advance development effort.

Multistep procedures are often used for the procurement of new systems which have defined performance objectives; but although the technologies that could accomplish the objectives are considered to be available, creative engineering effort is required to implement the technology. It is therefore essential that project managers be familiar with the types of multistep procurement procedures.

The two-step procedure is as follows.

Step One of the Two-step Procedure

1. An RFTP (request for technical proposal), which is based on the performance specification, is made to qualified companies.

2. Subsequent to receipt and review of technical proposals, technical discussions are held with those offerors whose proposed design approaches are feasible and responsive to the specification requirements. The primary intent of the technical discussions is to bring all proposals that are under consideration to the same level of competence in meeting the performance requirements of the specification. In addition, offerors whose proposals are evaluated as unacceptable are advised of their position and are told that they will not be included in the second step of the proceedings.

Step Two of the Two-step Procedure

1. The second step of the procurement is the solicitation for bids. Award of a fixed-price contract is made to the offeror submitting the lowest bid.

2. The offeror's technical proposal, upon which the successful bid is based, is one of the contractual documents and in some cases can take precedence over the performance specification.

The two-step procurement is used when the following circumstances exist.

1. There is insufficient technical data to prepare a design specification, as a result of which offerors must prepare and submit technical proposals. (This cir-

cumstance is also the basis for a negotiated procurement, which generally leads to an incentive or cost-reimbursable contract.)

2. A fixed-price contract is desired.

3. At least two offerors participate in the procurement.

The four-step source selection procedure represents an extension of the two-step procedure and provides for negotiations and discussion of all procurement elements, including cost, technical approaches, and other proposed contract terms. As with two-step procedures, it is essential that the project manager possess a comprehensive knowledge of each step of the procurement procedure in order to effectively implement the lead role throughtout the complete procurement cycle. The different phases of the four-step procedure can be summarized as follows.

Step One of the Four-step Procedure

1. Offerors submit documents called for in the TPR (technical proposal requirements), minus the cost information.

2. The procuring activity reviews and identifies in each proposal those areas which require clarification and further information.

3. Discussions are held with each offeror regarding any areas in question. (It should be noted that representatives of the procuring activity are precluded from revealing areas of deficiency or rendering any opinions on the merits of the offerors' proposals).

Step Two of the Four-step Procedure

1. Clarifications and other additional information that each offeror deems appropriate are provided as a result of discussions in step one. Cost proposals are submitted.

2. The procuring activity reviews offeror submissions and prepares preliminary PERs (proposal evaluation reports).

3. A determination is made as to whether any offeror's proposal falls outside of the competitive range and must therefore be judged unacceptable.

4. Discussions on clarifications and costs are conducted with offerors whose proposals are within the competitive range.

5. The procuring activity reviews the PERs and the competitive range to reflect the second round of discussions.

6. Final cost proposals are requested from the remaining offerors.

Step Three of the Four-step Procedure

1. Final proposals are received and evaluated.

2. One offeror is selected for negotiations to establish a definitive contract.

Step Four of the Four-step Procedure

1. Negotiations are held with the offeror selected in step three.
2. A contract award is made. (In the event negotiations break down, the second-choice offeror is invited to negotiate a contract award).

One significant difference between the two-step and four-step procedures is that in the two-step procedure, cost is not negotiable, whereas in the four-step procedure, cost is negotiable. An offeror that submits a submarginal technical proposal that is judged to be susceptible to being made acceptable is provided an opportunity to make its proposal acceptable during the step-one discussion of the two-step procurement. In the four-step procedure, both the cost and revised technical proposals are evaluated by the procuring activity to identify the propsoals that fall within the competitive range.

The competitive range for the technically acceptable proposals in four-step procurements is determined by the contracting officer of the procuring activity. The range is based on cost, technical responsiveness, and other factors. It should be noted that if the proposal of any offeror is evaluated as not technically acceptable, the proposal will be judged to be neither within the competitive range nor susceptible to being made acceptable.

7.5 Cost-type Contracts

The cost-type contract imposes the greatest risk on the customer and is used primarily in procurements involving significant R&D effort. Because of the risk imposed on the customer, it is essential that the contractor selected be established and have a record of proven integrity and reputation. Thus it is essential that the project manager effectively transmit the company's credentials. Inherent in the cost-type contract are the contractor's good intentions and best efforts in completing the project. If the all-important element of contractor integrity is deficient or lacking, then the customer will suffer, since the cost-type contract does not offer any monetary incentive for efficient operation.

There are many versions of the cost-type contract, but the two most widely used are the cost-plus-fixed-fee (CPFF) and the cost-plus-incentive-fee (CPIF). These are the cost-type contracts used for procurements from a commercial contractor as contrasted with procurements from a nonprofit organization, such as a university.

The CPFF contract establishes a fixed fee or profit based on some percentage of an anticipated contract cost negotiated by the customer and contractor and stipulated in the contract. The fee usually is about 7 percent of the stipulated contract cost. The contractor is reimbursed for all allowable costs that are incurred in completing the contract. Thus, if the contract cost is $100,000 with a fee of $7,000, and the contractor incurs costs of $200,000, the fee should remain constant. There are situations when the contractor is entitled to a fee on costs which exceed the contract cost. The more prevalent of such situations are as follows.

1. Change in scope of contract—e.g., quantity of items and specification changes

2. Costs incurred which are beyond the control of the contractor—e.g., strikes, insurrection, and forces of nature

3. Adverse effects on the contract due to the failure of customers to meet their contractual commitments—e.g., failure to deliver data and/or parts as required, late approval, or failure to make comments on design reports

The point that should be made is that contractors can obtain reimbursement for fees if excessive costs were incurred due to circumstances beyond their control.

Claims for fee on CPFF contracts go far beyond the amount of fee involved. The performance of a company is judged on whether the acceptable products are delivered on schedule and without cost overrun. Therefore, most contractors are anxious to maintain a good reputation by minimizing or eliminating the stigma of experiencing cost overruns. A cost overrun represents an amount by which the accumulated costs exceed the stipulated cost of the contract. Because such costs are unexcused (due completely to the fault of the contractor), they indicate how well the contractor's project manager has performed in managing the contract. A chronic history of cost overruns by any contractor is a very good indication that the contractor has a poor manager or is, in effect, in the wrong business.

The CPFF contract is used for procurements which involve R&D or for delivery of a first production item. The justification for its use is the fact that the costs (engineering, programming, manufacturing, and materials) cannot be estimated with any degree of accuracy at the time of award. In many cases, when a performance specification forms the basis of the procurement, a CPFF contract is used. Because of the risk that the customer assumes, the CPFF contract usually incorporates different types of clauses, which serve to enable the customer to exert close control over the performance of the contract. Some of the clauses include the following provisions: the customer approves major subcontracts, the customer obtains the title of property prior to delivery, allowable costs are defined, and overhead rates are negotiated.

The CPIF contract combines some of the features of the incentive and cost-type contracts discussed previously. This type of contract provides for the reimbursement of all allowable costs incurred by the contractor plus a fee based on an incentive formula for performance. It is generally used as a compromise when the offeror is unwilling to enter into a fixed-price contract and when the customer does not want to have a CPFF contract.

One of the various types of contracts described above will be selected by the

customer as the most feasible for a contemplated procurement. It must be recognized that the type of contract the customer would desire (e.g., a fixed-price
contract) is not always feasible, since it could be very possible that none of the
qualified prospective contractors would be interested or willing to enter into an
agreement which, although desirable for the customer, is too risky for the contractor. Therefore, a customer must analyze the contemplated procurement
from many points of view and make a determination as to which type of contract is most favorable and at the same time would be acceptable to qualified
companies interested in the type of procurement contemplated. When customers have some serious doubt as to whether their choice of contract types
would stimulate an acceptable response from interested companies, an alternative type of contract is solicited. The primary objective is to select a type of
contract that will promote competition for the procurement.

7.6   Analysis of Risk

The primary reason for the existence of any commercial organization is to make
a profit or, as a minimum, not to suffer a loss. In certain special and unique
circumstances, a company may knowingly enter into a contract which will
result in a loss to the company, but such action is generally taken on the premise
that the long-term results will be profitable. It is the responsibility of the project
manager to appraise and quantify the risk for consideration by top corporate
echelons in making final decisions. Some examples of these unique circumstances are as follows.

1. The company will purposely bid a price that is lower than its projected
costs because of knowledge or expectation of procurement of additional units
in the future. The company will be willing to sustain the loss on the first contract
in return for the engineering and production experience that would place the
company in a favored position for the future procurement.

2. The company is willing to enter into a contract which will result in a loss
in order to keep the key people of the working force employed during a slack
period. Since certain fixed expenses are incurred regardless of the amount of
business the company is doing, the amount of the loss is usually equivalent to
the prorated amount of the fixed expenses.

However, in its normal course of doing business, a company will enter into a
contract only if potential profit justifies the risk that the contract poses. Project
managers must address themselves to the following two basic questions as
regards the risk that their company will assume.

1. Can the company deliver the equipment within the time frame specified
in the contract?

2. Can the company design, manufacture, and deliver the equipment specified without exceeding the projected costs on which the contract price is based?

In practically any endeavor, time is money. In the case of a contract, if unforeseen complications or mistakes require the expenditure of more time than was planned for any phase, additional salary expenses, overheads, and other extra costs must be charged to the contract. If a cost-type contract is involved, the customer will provide for the additional expenditures. If a fixed-price type contract is used, the contractor would be required to absorb the additional costs. Therefore project managers involved in proposing and/or negotiating a contract must recognize the risk areas that exist in the project and provide whatever contingencies they judge are necessary in order to protect themselves. At the same time, they must recognize that they are involved in a competitive procurement and that they may price themselves out of consideration if excessive contingency factors were included in their price quotation.

In analyzing the magnitude of maximum or conservative contingency factors that would be appropriate for the RLM simulator system, the project manager would compile factors noted in Table 7.1. At the same time, the manager must appraise the competition on the procurement and recognize that application of the conservative factors could inflate the cost beyond the competitive range and price the company out of the procurement competition. On the premise that the basic cost estimates are competitive and that other offerors would apply their own contingency factors as was appropriate for them, the project manager makes judgments as to how to protect the Creative Electronics Corporation against unforeseen but statistically probable problems and at the same time produce a cost that would be competitive. In Table 7.1, the contingency factors that would be dictated by the competitive environment are listed.

It should be noted that for CPFF projects, the procuring activity assumes the risk of cost escalations, so that if any contingency factors are included by an offeror, they would be based on the desire to protect the rate of profit return.

The rationale upon which the contingency factors are based is as follows.

Engineering

1. *Breadboarding:*   The breadboard by its nature is a trial-and-error operation by which an operating model of a particular system design is verified. In the procurement discussed, the breadboarding would involve such systems as the FSS, optical, readout, and similar systems.

2. *Detail design:*   Although the detail design is based on what has been proved by the breadboard, some redesign effort required as an outcome of the test and checkout results should be recognized.

3. *Drafting:*   The drafting effort is directly related to the detail design effort and bears a proportionate contingency factor.

4. *Testing and checkout:*   This effort involves somewhat the same type of effort as the breadboard except that revisions or changes involve changes in the detail design and drafting effort.

TABLE 7.1   Contingencies for the Fixed-price RLM Simulator Procurement

Area of effort	Maximum or conservative factor, %	Competitive factor, %
System analysis	7	3
System design	10	5
Programming	10	5
Coding	7	3
Drafting	2	1
Manufacturing and fabrication	5	2
Integration	12	8
Testing	12	8
Shipping, installation, and acceptance	5	2

Programming and Coding

1. *Software analysis and interface design:*   The objective of this effort is to translate the system requirements into the appropriate software models and design approach. Reviews and verification procedures usually require significant revisions and repetition of verification procedures.

2. *Software detail design and documentation:*   This phase, an extension of the analysis effort involving extensive design verification efforts, requires revision and a repetition of the verification procedure.

3. *Coding and element testing:*   The coding involves the translation of the software design into a program that requires testing and verification. Errors which are statistically unavoidable require analysis and troubleshooting procedures.

4. *Integration and testing:*   The integration and testing of the hardware-software system is generally a lengthy, detailed procedure involving the recording of data and correction of malfunctions.

Manufacturing

The FPI contract requires that contingencies be provided to cover fabrication, wiring, assembly, and other costs resulting from design revisions and changes, excessive spoilage, and other similar costs.

Materials

The contingencies for materials cover costs for components, raw materials, and subcontracted assemblies that might be required over and above those costs normally anticipated because of redesign and other adverse factors.

The magnitude of the contingency factors would, to a large degree, be derived from the statistical records of costs incurred on previous contracts that were due to mistakes and errors and could be expected on the procurement at hand. For instance, the statistics on drafting might show that 5 percent of drafting hours involved correcting errors uncovered by the checker; therefore an addition of 5 percent to the estimated number of drafting hours would be considered normal and reasonable.

If project managers provide for the maximum contingency factor, they may price themselves out of consideration for the procurement. They therefore must use their knowledge of the type of effort involved, the past performance of their company team, and the competition they face to arrive at a competitive contingency figure.

7.7 Summary

The two basic types of contracts are the cost-reimbursable contract and the fixed-price contract, both of which are implemented by formal advertising and negotiation techniques as applicable. There are many variations of each type, and each variation is best suited for a particular type of procurement.

The cost-reimbursable type of contract is used for procurements in which the effort and costs required to complete the contract cannot be accurately estimated. Such contracts include those requiring significant R&D effort. Because the contractor is reimbursed for all allowable costs incurred on the procurement, there is little risk associated with this type of contract.

The fixed-price contract is one in which the contractor must absorb all or part of the incurred costs which exceed the contract target price. By the same token, the contractor is "rewarded" for all or part of the amount if the incurred costs

Risk on Contractor – Fixed Price				Risk on Customer – Cost Type			
Firm fixed-price	Fixed-price with escalation	Fixed-price with redetermination	Fixed-price Incentive	Cost	Cost-plus fixed-fee	Cost-plus-Incentive fee	Time and material
Degree of Risk				Degree of Risk			
Where Applicable							
1.Design specification 2.Competition 3.No negotiations	1.Design specification 2.Escalating labor and material costs 3.Negotiated	1.Quantities 2.Long delivery 3.Fluctuating labor and material costs 4.Specification not definitive 5.Negotiated	1.Performance specification 2.Negotiated 3.Fluctuating labor and material costs 4.Risks	1.Performance specification 2.Research and development 3.Negotiated	1. Performance specification 2.Development 3.Negotiated	1. Performance specification 2.Development 3.Negotiated 4.Possible cost reduction	1.Performance specification 2.Emergency 3.General services

FIG. 7.1 Types of contracts based on risk.

are less than the contract target price. Thus the contractor assumes a greater risk on the fixed-price type of contract. Because there are different risk factors inherent in each basic type of contract, the company bidding on a procurement must evaluate such risks and adjust the bid price accordingly.

Multistep procurement procedures are used for projects which cannot be supported by a definitive specification that would be used for an advertised procurement, but which do not involve research or significant development effort. The most frequently used procedures are the two- and four-step procedures. The multistep approach constitutes a hybrid of the formal advertised and the negotiated procurement procedures.

Figure 7.1 summarizes the degree of risk that the contractor and customer assumes with the different types of contracts.

BIBLIOGRAPHY

Defense Contract Management for Technical Personnel, Naval Materiel Command, Washington, D.C., 1980.

Defense Acquisition Regulation 4-107, U.S. Department of Defense, Washington, D.C., 1978.

8

CONTRACT
CLAUSES

8.1 Definition

Any commercial corporation or government agency has procurement policies which are expressed as standard, or boiler plate, contract clauses. These clauses usually exist as printed pages and are attached to the particular applicable contract.

Since the boiler plate clauses set forth the operating ground rules for a procurement and describe the rights and obligations of all parties to a contract, it is highly important that the project manager know the details for all their terms and how they affect the procurement under consideration. A brief description of the clauses with which project managers must be concerned and how they apply to the procurement of the RLS will be given. The clauses to be discussed are typical of those used by the government for procurement contracts but are also applicable, to a large extent, in commercial-type procurements.

It should be noted that if any clauses of the contract schedule and boiler plate are in conflict, the language cited in the contract schedule takes precedence.

8.2 Changes Clause

This clause is applicable to both the cost-reimbursable and the fixed-price type of contract, and the wording is essentially the same. By the terms of the changes clause, the customer may, without notice, make certain changes in the contract requirements. The types of changes that fall within the domain of the changes clause are:

1. Changes in design, drawing, or specification for special equipment being created for the user.
2. Changes in the method of packing or shipment or in destination.

If the changes affect the cost, delivery, or other areas of the contract, the contractor must make a written claim, which serves to initiate a review, a negotiation, and an equitable resolution of the claim. When a change requires additional funds, it is presumed that the customer, be it the government or a private organization, has budgeted adequate money to pay for the change.

When procurements are under consideration, the project manager must not lose sight of the fact that customers can exercise their rights in the changes clause at any time and that the contractor has a primary obligation to comply. Whereas it would be unreasonable to expect the project manager to possess clairvoyant powers and predict the changes that will be issued (although in some cases intimate knowledge of the procurement and very close contact of the manager and the marketing staff with procurement officials can provide information that allows changes to be anticipated), the project manager should ensure that the design is as flexible as possible so as to permit any future changes.

Changes can effect what amounts to a decrease in scope. Decreases in scope resulting from the application of the changes clause are relatively rare, but the same principles and procedures as those discussed apply in reverse.

The changes clause of a contract is intended to give the customer flexibility in satisfying the equipment requirements and to provide for improved features or performance characteristics which were not envisioned at the time the contract specification was adopted. The clause is not to be viewed by the customer's project team as a convenient tool for rectifying errors that result from a poorly conceived and written specification.

Changes to a contract, particularly a government contract, can assume the form of verbal or informal directives. Often a change to a contract may be unintentionally directed and can end up as being contractually binding. Changes of this type are identified as constructive changes and are discussed in Chapter 10.

8.3 Allowable Costs

The boiler plate clauses for cost-reimbursable contracts provide for controls and approval by the customer of the costs that are accumulated. In general, only those costs incurred in connection with the contract will be allowed and reimbursed. The criteria for determining what constitutes an allowable cost in government contracts are defined in the procurement regulation documents. In contracts between private parties, the terms of the contract generally define what constitutes an allowable cost.

An allowable cost is generally defined as one which would be considered reasonable for and applicable to the contract in question. A cost which would be incurred by a prudent person as a result of sound business practices would be considered reasonable. If, for instance, the contractor purchased subassemblies from another company as a noncompetitive subcontract and the subcontract costs were judged to be unreasonably high, those costs might be judged as not allowable by the customer's auditor. Discussions between the contractor and customer would be required to establish mutually agreeable figures, if possible.

An allowable cost must be applicable to the contract being changed. Engineers and other employees must charge their time to the project on which they are involved. Therefore, the contractor must have a system which will provide an accurate means of charging costs to the appropriate project.

The allowable-cost clause also sets forth the controls on the payment of profit or fee and the schedules of payment of a cost-reimbursable contract. The contractor is obligated to advise the customer at a predetermined time of any indication that the cost stipulated in the contract will be exceeded. In no case will the contractor be reimbursed for costs exceeding those in the contract unless they are authorized by the customer as a contractual price change.

8.4 Inspection and Correction of Defects

With both fixed-price and cost-reimbursable contracts, the contractor is obligated to correct any defects uncovered prior to acceptance. In addition, the contractor is usually required to correct any defects uncovered up to 6 months after acceptance of the equipment. The contractor is required to provide "reasonable" facilities and assistance to the customer when the equipment is being tested. The customer usually has authorization to perform whatever tests or inspections are deemed practical prior to the completion of acceptance tests.

With the cost-reimbursable contract, the contractor will receive payment for direct costs incurred as specified under the allowable-cost clause, although payment of fee will not usually be allowed at that time. The exception to the reimbursement of costs in the case of uncovered defects is when fraud or similar practices on the part of the contractor are revealed.

With the fixed-price contract, the contractor must remedy defects uncovered up to 6 months after acceptance, with no reimbursement. If the contractor fails to complete the necessary corrections, an equitable sum can be withheld by the customer. Any disagreements are handled in accordance with the terms of the disputes clause.

In preparing the cost proposal for a procurement, the project manager must provide for the time, facilities, and costs that will be required for the inspection and the corrections that would be necessary under the proposed contract.

8.5 Subcontracts Clause

About the only restriction placed on subcontracting on a fixed-price contract is that wherever possible, subcontracts must be awarded on the basis of competitive bidding.

In the case of cost-reimbursable contracts, the subcontracts clause affords the customer some very definite controls regarding the subcontract awards and effort. By the terms of this clause, the contractor must advise the customer of contemplated subcontracts. Contingent upon the size and type of subcontract that is under consideration, approval of major subcontracts must be received from the customer. However, consent of the customer to use a particular subcontractor does not relieve the contractor of the responsibility for meeting the terms of the contract in the event the subcontractor fails to perform.

In the event of any contractual difficulty, such as slippages, claims, or other disagreements, the subcontractor is obligated to report such difficulties to the prime contractor as soon as possible.

It has been indicated that Creative Electronics would probably subcontract the detector and other systems. It will be the responsibility of the project manager to draft a set of specifications, solicit subcontract proposals, and participate in the negotiation of the best possible contract. It would be in the interests of the company to enter into a fixed-price subcontract with its vendors.

8.6 Termination Clause

The two basic types of contract terminations that can be applied to both fixed-price and cost-reimbursable contracts are: termination at the convenience of the customer or government and termination due to contractor default.

A convenience termination is one in which customers decide that the material under procurement is no longer required and that they are prepared to assume whatever losses are associated with termination. In cases of convenience terminations, the contractor ceases all work and issues cancellations on all purchase orders and subcontracts immediately upon receipt of the termination notice. Within a specified time (usually a year) the contractor must submit the termination claim. The termination claim includes all incurred costs, including cancellation charges on orders and subcontracts, special expenses associated with the termination effort, and other similar costs.

With the cost-reimbursable contract, the costs that are claimed are verified as applicable, and the fee is negotiated. If an impasse regarding claimed costs or fee develops between the contractor and customer, the matter is referred to a higher authority for resolution (legal courts for commercial customers and the Settlement Review Board if the customer is the government).

With a fixed-price contract, legitimate costs and fees will be determined and negotiated when a convenience termination is executed. If the evidence of a procurement indicates that the contractor would have exceeded the ceiling contract price (thereby experiencing a loss on the contract), a sharing of the loss by the customer and contractor is negotiated. Again, the customer has inherent rights of appeal, as cited above.

Termination by default is applied in contracts because of the failure of the contractor to deliver the item or items under procurement as specified or because there is concrete evidence of the failure of the contractor to make progress in the program, thereby supporting the supposition that the terms of the contract will not be met. The process in such terminations involves officially notifying contractors of their lack of progress, which in turn requires a response within a stipulated number of days. If the response is not made or is unsatisfactory, the termination action can be executed.

In cases of default terminations of cost-reimbursable contracts, all substantiated costs are reimbursed. However, fee will be paid only on those items that have been accepted. Thus, for an item of the contract which is 90 percent complete, only those costs associated with the item will be paid with no fee.

The procedures and reasons for default termination on fixed-price contracts are the same as those discussed for cost-reimbursable contracts. However, the contractor will be paid only for those items that have been accepted. In the case of the above-noted item that was 90 percent complete, the contractor in a fixed-price contract will receive no reimbursement for the item. Thus the contractor stands to experience a complete loss on any outstanding items that have not been delivered. In addition, the customer has the right to procure the unaccepted items from another source, and the defaulted contractor is then legally obligated to bear any costs incurred by the customer that are over and above the original costs of the items involved.

8.7 Delays

When a contractor becomes delinquent in the execution of the contract or in the delivery of items on schedule, the customer is obliged to promptly indicate that some type of action will be taken or it will jeopardize its rights in the procurement. For government acquisitions, the procedures are clearly defined and mandatory for the procuring activity. For commercial acquisitions, prudent action indicates that similar procedures should be followed.

In situations when delays are encountered, the procuring party normally initiates one of the following types of action.

1. It negotiates an extension in the delivery schedule.
2. It terminates the contract owing to default.
3. It continues with the contract in the delinquent state but establishes that it does not waive its rights.
4. It negotiates a no-cost termination.

It is incumbent upon the project manager to work for the alternative that best serves the organization when the procurement contract is in a delinquent state.

Delays are excusable if caused by circumstances beyond the control and without the fault or negligence of the contractor. Such causes would include fires, floods, and other similar causes that are identified as "Acts of God." Strikes, insurrection, and other human-created problems would also be identified as excusable delays.

The same type of excusable delays apply to subcontracts on a project, providing the following additional conditions are met.

1. The contractor diligently monitored the subcontract to establish that the difficulties existed or were developing.

2. The contractor diligently attempted to seek alternate sources for the subcontracted items but was unsuccessful.

3. The subcontractor did not contribute to the cause of the delay either by action or by inaction.

A delay may be considered excusable if caused by the failure of the customer to execute some required contractual obligation. In such a case, the contractor may be entitled to receive reimbursement for expenditures due to the delay.

8.8 Disputes Clause

The same disputes clause language and terms are applicable to both the cost-reimbursable and the fixed-price contract. Any dispute (other than one involving allowable costs) not resolved by agreement is arbitrarily decided by written notification by the customer. The clause concerns questions of fact. After the decision is rendered by the customer, the contractor, within a specified time, can appeal. In the case of government contracts, the appeal is made to the Secretary of Defense (Navy, etc.) within 30 days. For commercial procurements, the appeal can take the form of legal action in an appropriate court of law.

If the dispute involves issue of allowable cost on government contracts, the contractor can appeal to the government auditor within 60 days after the determination by the government. Adverse decisions can be further appealed to the cognizant cabinet secretary.

For commercial procurements, the contractor can appeal to courts of law.

For government procurements, the relief that a contractor can seek is well defined by the universally applied disputes clause. For commercial contracts, the contracting parties should agree to procedures for resolving disputes and incorporate the specific agreement in the contract. Pending final resolution of any dispute, contractors have the obligation to pursue in a diligent manner the execution of their contractual obligations. This is specifically stated in the clause for government contracts.

8.9   Customer- or Government-furnished Property

The same conditions relative to customer-furnished property are applicable to cost-reimbursable or fixed-price contracts. The customer is contractually obligated to deliver the specific items by the dates set forth in the contract or, in lieu of specific dates, at a time sufficiently early to permit contractors to utilize the property in time to fulfill their obligation. If the property is not furnished as required, the contractor must serve written notice of the delinquency so that an equitable adjustment in cost and delivery can be made in accordance with the provisions cited in the changes clause. In addition to the responsibility that the customer has for timely delivery, the property must be suitable for use in the contract. It should be added that the customer-furnished property can include data, reports, drawings, and other materials as well as equipment.

Title to the property remains with the customer; however, the contractor has the contractual responsibility to maintain, repair, and preserve the property in accordance with sound commercial practices. If major failure or damage to customer-furnished property occurs which is not due to any negligence on the part of the contractor, the customer is obligated to make the necessary repairs or replacement and is liable for the responsibility of any schedule slippages or incurred costs that may result.

Property furnished by the customer may be agreed upon to be in "as is" condition. In effect, in such an arrangement the contractor agrees to utilize equipment for the contract that may not be in a suitable condition for its intended use. The contractor then assumes the responsibility for making the repairs, etc., necessary in order to provide equipment in a suitable condition. Generally, the contractor also assumes the responsibility for making any repairs or performing any maintenance that may be requested.

8.10   Patent and Copyright Infringement

The patent and copyright infringement clause is applicable to both the cost-reimbursable and the fixed-price type of contract. In the event that the contractor, in fulfilling the contract, infringes on the rights assigned to a third party, and such party initiates infringement litigation, the contractor is obligated to immediately file notice with the customer of the litigation. The customer is liable for all infringement charges arising out of the performance of one of its contracts. The customer is also responsible for reimbursing the contractor for any evidence that is requested except in those cases in which the contractor has agreed to indemnify the customer against the claim being asserted.

When a contractor indemnifies the customer against infringement, the contractor is agreeing to assume litigation and possible penalties for any patent infringement that may occur during the execution of the contract. The general conditions associated with such indemnity include the following: the customer must notify the contractor of any infringement allegations, the contractor must be given adequate time to defend the allegation, and the infringement must not be the result of the customer's direction.

8.11 Filing of Patents

The clause relating to the filing of patents is applicable to both types of contract. The contractor may pursue a patent application for an invention developed under a customer or government contract. However, the contractor must grant to the customer an "irrevocable, nonexclusive, and royalty-free license" to use the patent as desired. In cases of government contracts, the contractor retains the right to use the patent or sell patent rights to other parties for use on commercial applications. Even if an invention has no commercial value, a company may still desire to expend the funds to procure a patent to serve as a prestige asset or to enable the company to dominate its particular field.

If contractors do not desire to obtain a patent on an invention developed on a government contract, they still are obliged to furnish the government with all the necessary information and disclosures so that the government may apply for a patent. In such a case, the patent would be assigned to the government. If contractors resist submitting disclosures to the government on developments deemed to be subject for patents, the government can withhold a specified sum of money in payment of the contract.

8.12 Overtime and Shift Premiums

Overtime and shift premiums are not allowable on cost-reimbursable contracts unless specific written approval for such premium time is received by the contractor from the customer. Overtime and shift premium payments are extended if they are:

1. Required for emergencies due to accidents, equipment breakdowns, etc.
2. Required for indirect labor employees performing project management, administration, maintenance, etc.
3. Required to perform tests and functions that cannot be interrupted.
4. Required to lower overall costs to the customer.

For fixed-price contracts, the only restriction is that the customer reimburse employees working in excess of 8 hours per day.

8.13 Value Engineering

The objective of a value engineering clause in a contract is to provide an incentive for the contractor to analyze the contract requirements for the equipment

TABLE 8.1 Summary of Significant Boiler Plate Clauses for Cost-reimbursable and Fixed-price Contracts

Type of clause	Cost-reimbursable contract	Fixed-price contract
Changes clause	Unilateral changes within scope of contract can be made without notice by customer. If change affects scope of contract, equitable adjustment in contract is made.	Same
Allowable costs	Incurred costs are subject to the approval of customer. Contract cost cannot be exceeded unless specifically authorized.	Not applicable
Inspection and correction of defects	Contractor provides facilities and corrective effort up to 6 months after acceptance. Costs in general are reimbursed.	Contractor provides facilities and corrective effort up to 6 months after acceptance. Costs not reimbursable.
Subcontract clause	Approval by customer required for specific types and sizes of subcontracts.	Competitive bids required.
Terminations convenience	Notice served by customer. Contractor submits termination claims.	Same. If contract would have exceeded ceiling cost, loss is assumed by contractor.
Default	Issued when lack of progress is shown or when contractor fails in contractual efforts. Notice is issued to contractor and response required in specified time.	Same
	Contractor reimbursed for costs incurred. Fee paid only for accepted items	Contractor reimbursed only for accepted items with fee. Contractor not reimbursed for any costs involving unaccepted items.
Excusable delays	Delays beyond the control of the contractor are excusable.	Same. If delays caused by customer, contractor is reimbursed.

Disputes clause	If agreement not reached, customer renders unilateral decision which can be appealed. Disputes involving questions of fact and allowable costs are handled and appealed differently. Contractor obliged to perform pending appeal.	Same
Government-furnished property	Government is obligated to deliver property as required in contract in suitable operating condition and is responsible for repair or replacement if major difficulties occur. Adverse effects on contracts due to failure to meet property requirements subject to reimbursement or rescheduling. Contractor responsible for normal maintenance and repair.	Same
Patent infringement	Government liable for infringement of patents by contractor in performing under a contract. Contractor is reimbursed for information furnished by contractor for use in defending suit.	Same
Filing of patents	Contractor may file but must assign all rights to government. Contractor can use patents in commercial application. If government files in lieu of contractor, contractor must furnish all disclosures.	Same
Overtime and premium shifts	Written permission required for utilization. Valid only in special emergency or circumstances that preclude standard 8-hour operation per day.	No restrictions, provided employee is reimbursed.
Value engineering	Contractor may submit proposals if value engineering clause is included in contract.	Same

under procurement and to solicit proposals for changes which will result in cost savings. The incentive is that if the value engineering proposal is accepted, the contractor will share in the cost savings resulting from the proposed change.

The basic conditions that are necessary to support a value engineering reward include the following.

1. The contract must include special provisions to permit the implementation of value engineering proposals.

2. The proposal must address a specific contract requirement that can be modified at a savings in cost without adversely affecting the quality and performance capability of the product under procurement.

3. The value engineering proposal must be accepted by the customer.

4. A specification and contract amendment to reflect the cost savings cited in the value engineering proposal must be executed.

Value engineering concepts are mostly applicable to procurements based on design specifications involving larger quantities of items. For instance, the Navy at one time required fixtures such as signal lights used aboard ships to be fabricated out of bronze and machined to close tolerances. Some time ago, a contractor for the item proposed an aluminum alloy material that was to be cast to tolerances that were not as precise as specified. The value engineering proposal included the following basic justifications for the specification deviations.

1. The close tolerances specified for the product were unnecessary for the product under procurement.

2. The substitution of the aluminum alloy for the bronze housing could be accomplished without in any way affecting the performance, life cycle, maintenance requirements, or any other requirements of the product.

After evaluating the value engineering proposal, the government accepted the proposal, revised the specification language, amended the contract, and shared with the contractor the savings realized by the changes that were implemented.

Value engineering contract clauses are generally not applicable to procurements based on performance specifications, such as the RLM simulator. In effect, the project manager performed what amounted to a value engineering analysis in considering the alternative approaches for the technical proposal.

8.14 Other Common Clauses

In addition to the above major standard clauses, there are numerous other clauses in the boiler plate. Most of the other clauses cover areas which do not relate directly to the project manager's functions, but their existence should be noted. Some of these fringe clauses cover the following areas and are handled by different departments of the contractor.

1. Buy America Act
2. Convict labor
3. Payment of overtime and shift premiums
4. Nondiscrimination in employment
5. Insurance liability of third parties

8.15 Summary

Since a primary responsibility of the project manager is to meet all of the requirements of the contract, including cost and profit objectives, a comprehensive knowledge of the contract terms is essential. The standard or boiler plate clauses can be a significant factor in frustrating the achievement of the project objectives if they are not properly applied. Table 8.1 provides a summary of the most significant types of boiler plate contract clauses utilized for government procurements and gives a brief description of each type as it would apply to cost-reimbursable and fixed-price contracts.

BIBLIOGRAPHY

Defense Contract Management for Technical Personnel, Naval Materiel Command, Washington, D.C., 1980.

9

NEGOTIATIONS

9.1 Definition of Negotiations

Subsequent to submitting the proposals on a procurement, the project manager must prepare for the negotiations that may take place after the initial proposal evaluations by the procuring activity have been completed. Whereas in the past, negotiations may have related primarily to price bargaining, discussions on contemporary procurements by government and industry involve scheduling, performance terms, and other contract provisions in addition to price. The objective of the negotiations is for the parties to reach a meeting of the minds and to document as contractual obligations the terms on which agreement was reached. Another way in which a negotiation can be described is as follows.

Procurement by negotiation is the art of arriving at a common understanding through bargaining on the essentials of a contract, such as delivery, specifications, prices, and terms. Because of the interrelation of these factors with many others, negotiation is a difficult art and requires the exercise of judgement, tact, and common sense. The effective negotiator must be a real shopper, alive to the possibilities of bargaining with the seller. Only through an awareness of relative bargaining strengths can a negotiator know where to be firm or where to concede on prices or terms.

Not all procurements lend themselves to negotiations. On standard items in common use for which the costs are fairly well established, the buyer would award a contract to the lowest bidder. The project manager, however, is involved in procurements which often represent the first of a kind and for which significant creative engineering effort is necessary. For such projects, the project manager must be prepared to participate in rigorous procurement discussions.

Procurements that require the development of new and complex systems necessitate technical discussions to clarify and supplement information contained in the proposal. Such discussions are held to establish whether the technical proposal is acceptable or not acceptable. The term "technical negotiations" is a misnomer since the use of the work negotiations implies an acceptable technical proposal which has not as yet been established.

If a proposal is categorically evaluated as not acceptable due to significant deficiencies, it will not be involved in technical clarification discussions. However, proposals that are considered acceptable and proposals that the evaluators judge to be susceptible to being made acceptable because their deficiencies are potentially correctible will be included in technical discussions.

For two-step procurements, the first step is confined to the clarification of technical and other terms of the proposed contract. If a technical proposal is evaluated as not acceptable but susceptible to being made acceptable, the task of the project manager is to provide whatever information is necessary to make the proposal acceptable. However, revising the basic design approach is not permitted, since such a major change constitutes a new proposal. Subsequent to the first step, the firm price is submitted, as it would be for an advertised procurement.

Other types of procurement procedures, such as the four-step approach, involve the concept of the competitive range as well as the "susceptible to being made acceptable" notion. The competitive range includes the salient procurement factors, such as technical excellence, contract terms, delivery, and cost. In all discussions, the project manager plays a key role for the company and must treat each element of the process of clarification—amendments, supplementary inputs, and meetings with procurement activity officials—as part of the process that will, it is hoped, lead to a contract award.

9.2   Establishing Negotiation Parameters

The primary goal of the project manager is to obtain the contract at terms which favor his or her company to the greatest possible degree. (In matrix organizations, the functional manager would usually participate in the negotiation along with the project manager). In planning for the negotiation, which will involve the parameters of cost, delivery, etc., the project manager must establish the points beyond which the contract would not be acceptable. There are many factors which enter the picture in deriving the limits of acceptability of the contract. Some of these factors are portrayed in Figure 9.1, which provides a

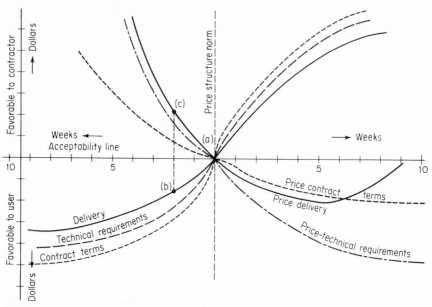

FIG. 9.1   Variations in price based on procurement revisions.

graphic display showing the relative effects that changes in delivery schedules, technical requirements, or contract terms would have on the cost of producing a product. The intersection of the acceptability line with the price structure norm (point *a*) represents the price that Creative Electronics proposes for the trainer as specified in the RFP. The acceptability line has been established on the basis of such considerations as competition, the desire of the bidder for the contract award, experience, and risk factors associated with the procurement. If during negotiations any changes to the procurement parameters are considered, the project manager can refer to a chart similar to Figure 9.1 and derive an instant guide as to what impact a suggested change in any of the major procurement provisions may have on the cost structure and whether the change is practical.

The increments shown on the vertical scale would be expressed in dollars over and under the bid price offered by the contractor when the proposal was submitted. A figure that is less than the proposed price would fall below the acceptability line and therefore be favorable to the user, since it represents a cost reduction on the proposal. A price above the line would be favorable to the contractor.

The horizontal acceptability line would be scaled to reflect the following: time, for use with delivery and price-delivery curves; degree of complexity, for use with technical requirements and price–technical requirements curves; and degree of contract hardship, for use with contract terms and price–contract terms curves.

If we were to apply the curves of Figure 9.1 to the RLM simulator procurement, point *a* would represent the position of Creative Electronics at the time the proposals are submitted. In the interest of clarification, assume the bid price at point *a* to be $500,000. Each horizontal increment above the acceptability line could represent an addition of $10,000 to the bid price, and each increment below the acceptability line could represent a subtraction of $10,000.

In like manner, the delivery requirements can be scaled to weeks along the acceptability line. Point *a* represents the delivery requirements of 130 weeks. Each horizontal increment could represent 2 weeks added to or subtracted from the 130-week schedule. To illustrate the application of Figure 9.1, suppose the customer wanted to accelerate the delivery schedule by 4 weeks and the consideration in this case is money. The project manager, in referring to the curves of Figure 9.1 as a guide, can establish from the price-delivery curve that the contract price should be increased by $40,000 to permit the delivery accelerations. The figure of $40,000 is derived from the estimated cost of overtime, risks associated with accelerated effort, and other factors of this nature. The points in question are designated *b* for the new delivery and *c* for the increased price. It should be noted that any point on the price-delivery and delivery curves must be on the same vertical line in order for the correlation to be valid and accurate.

Of particular interest is the slope of the two curves in question. The price-delivery curve slopes sharply upward when correlated with increasingly shorter delivery, as expressed by the delivery curve. In other words, if the customer wanted the delivery improved by 8 weeks, there would be no practical point on the price-delivery curve that falls on the same vertical line as the 8-week improved delivery point, which means that the improved delivery is impossible as far as contract price is concerned.

The slopes of the various curves are significant. For instance, in the favorable-to-user area, the delivery curve approaches a slope that is parallel to the acceptability line. The notion being conveyed is that beyond a certain point (approximately 12 weeks, in this instance) any shortening of delivery time for the trainer does not result in any increased benefit to the user, even if such schedule shortening were possible or economically feasible. The same conclusion can be drawn by noting the slopes of the technical requirements and contract terms curves in this area.

It should be noted that the slope of the price-delivery curve reverses itself as the 130-week delivery schedule is extended beyond 8 weeks. Thus, up to a point, the contract price can be reduced if the delivery requirement is relaxed or extended. At the 130 plus 8-week point, however, Creative Electronics anticipates that the RLM simulator will be ready for shipment, even as built at the leisurely schedule, and to keep the unit in the plant would result in the unnecessary accumulation of costs. It should be noted that the delivery curve slopes downward, indicating that beyond the 8-week extension, the delivery conditions cease to be in favor of Creative Electronics.

The curves representing contract terms and price-contract terms, as well as the curves representing technical requirements, would be applied by the project manager in a similar manner during negotiations.

The establishment of the price-delivery and delivery curves was relatively easy, since these two curves could be derived directly from numbers (such as numbers of weeks and dollars). However, in the case of the curve representing contract terms, the task is more difficult. How, for instance, could a more stringent requirement for reliability of operation be readily expressed in dollars except through a thorough design analysis? Since the curves are intended primarily as a guide to the project manager during negotiations, the contract terms curve could express estimated degrees of hardship that would be imposed on Creative Electronics, and the degrees of hardship could be correlated with the price-contract term curve. It is in this area that the technical knowledge, familiarity with the proposed contractual and technical details, and experience of the project manager play a major role. If during negotiations the proposed contract terms are revised, the project manager must be able to judge their impact on the effort of Creative Electronics and thereby relate such impact in terms of price.

Again, in Figure 9.1, the slopes of the contract term curve and the price-contract term curve indicate that, at some point, no amount of money could compensate Creative Electronics if the contract terms are made too severe. In like manner, there is a level at which the price-contract term curve becomes horizontal, indicating that no amount of relaxation of the contract terms would reduce the procurement price.

The slopes of the technical requirements and the price-technical requirements curves are essentially the same as those of the contract term and price-contract term curves. Again, the technical requirements curve must be expressed in some convenient scale, such as the degree of impact on the effort that is required in order to have the variations of the technical requirements expressed in the proposed contract price. As before, the project manager must have the knowledge and capability to judge any proposed revision of the technical requirement in terms of price increase or decrease during negotiations.

The variations that could evolve during negotiations have been discussed and presented in terms of the effects on the proposed contract price. In general, most negotiations would center around the question of how the price would be affected by changing some requirement. It is not uncommon for factors other than price to be the prime consideration. For instance, a family of curves which demonstrates how the delivery schedule might be varied can be drawn as different revisions in proposed contract requirements are suggested. If the technical requirements are tightened beyond a certain point, the delivery would be extended beyond a reasonable point and would be expressed by an upward sharp slope of the delivery-technical requirements curve. In other words, the technical requirements, such as accuracy and response speed, could be made so stringent that they could not be met regardless of the amount of time that might be extended to the contractor to accomplish the task.

The curves of Figure 9.1 illustrate one convenient tool which can serve the project manager in negotiating a contract. It should be emphasized, however, that in order to have the tools serve to the maximum advantage, their user must be thoroughly familiar with how they were derived and with the purpose for which they were designed. The reason for this is that for most of the curves the selection of a particular point is essentially the result of judgment.

9.3 Analysis of the Buyer's Position

A negotiation can be viewed as a contest in which parties will use to their advantage any situation that exists or that may develop during the proceedings. In any contest, one fundamental requirement is to study and learn everything possible about the opposition and its position.

In the case of the RLM simulator, the first obvious bit of information is that a specific requirement for the trainer exists. Further, the project manager should note that the radar to be simulated is the AN/APQ-28 Search Radar being designed for use on the P7K aircraft, which are scheduled for delivery to the fleet in the very near future. The AN/APQ-28 radar is known to be a complex unit requiring special skills for its operation which must be developed as soon as possible. Therefore the project manager can conclude that the simulator is required as soon as possible and that delivery is of prime importance.

In addition to the above, the project manager might note that the end of the fiscal year is approaching. In government procurements, money generally has to be obligated prior to the end of the fiscal year. Therefore, heavy pressure would be imposed on the procuring agency to enter into a contract as soon as possible. In addition to the fiscal year deadline, the buyer is limited by the amount of money that is budgeted for the procurement. For government procurements, budget information is restricted, but for commercial procurements such information is occasionally revealed and can serve as a guide to preparing the proposal bid.

If there are indications derived from contacts with the technical personnel of the procuring agency that a particular design approach is viewed with favor, project managers should be prepared to emphasize the advantages of their system if another design is favored or to reinforce the favorable views of the procuring engineers if their system is favored.

Another factor to use to advantage relates to the identity of the competitors trying for the contract award—information regarding their proposed approaches, their probable costs, and their past performance. Information regarding the general attitude of the procuring personnel towards each of the competing companies would also be of value. Again, this type of information must be derived from experience and familiarity with the competing companies. The identity of the competing companies can usually be obtained by observing who the participants in the bidders' conference are or by logically deducing who the competitors would be. The design approaches of competitors can be deduced from past associations and conversations with employees of those companies. It should be recognized that the fact that two companies are competitors does not preclude their associating with each other. In any general field of industry, there is a constant shifting of personnel among companies, and with the shifting there are always exchanges of information which the new employers record.

All of the available intelligence information can be correlated and organized to establish hypothetically what the customer's attitude might be toward a design that a competitor might be expected to offer.

In essence, it is incumbent upon the project manager to analyze the probable attitude of the buyer toward the competitors and to emphasize during negotiations the advantages that the Creative Engineering proposal offers over the competitive proposals without directly mentioning the competition.

9.4 Knowledge of the Product

It is mandatory that project managers be intimately familiar with the technical and cost details of the items under procurement and that they be prepared to discuss these points with confidence. They could have at their disposal selected personnel who are expert in the different important areas. If the negotiations were to require the inputs of one of the experts of the negotiation team, the project manager would act as the spokesperson after privately discussing the particular point under consideration.

In addition to being able to provide information on the technical aspects of the RLM simulator, the project manager must be able to discuss and answer questions relating to the construction, the types of materials, the components to be used, and other related aspects of the simulator.

The type of materials and components to be used in the simulator would be of concern to the user. For instance, since the simulator must be capable of uninterrupted operation, the project manager must be able to describe how the materials and components to be used will meet the reliability requirements. With the digital computer approach proposed by Creative Engineering, the quality of the computer program would be of concern to the procurement authorities. Therefore, it would be desirable to supplement the proposal information with additional data describing how the software is developed, tested, verified, and integrated with the hardware so as to make sure that the system will operate with the desired fidelity and reliability.

An area of the procurement which has been touched upon lightly and which is important to the user relates to the side items. Among these side items, manuals are particularly troublesome to most users, and as a result there is considerable discussion as to how the manuals are to be prepared, who will prepare them, what their content will be, etc.. The project manager should be briefed in this area and should be prepared to discuss the side items in detail sufficient to satisfy the prospective user.

9.5   Knowledge of Cost Figures

The cost of a procurement is generally of primary concern to a prospective user, and the negotiations for such types of procurements, will be primarily concerned with the cost structure of the proposed contract. The amount of detailed discussion and the extent to which a cost breakdown is explored are dependent on the type of contract that is solicited. In a CPFF type of contract, in which the contractor is reimbursed for all allowed incurred costs, the direct cost proposed is more or less academic and a matter of audit. The pertinent issue is an evaluation of how efficiently a particular company can perform. For instance, if company A bid $100,000 on a job but past experience indicates a probable cost overrun of $75,000, then it would be judicious to make an award to company B, which bid $125,000 but is evaluated from past experience as having a realistic price. Thus, in a CPFF procurement, the issue of direct cost is secondary, and detailed cost negotiations will generally be concerned primarily with overhead rates and profit.

On the other hand, the cost negotiations on fixed-price procurements can be expected to include a detailed explanation of how cost figures were derived and a justification for those cost figures in any particular area. Project managers must be cognizant of the extent of effort, the type of effort, contingency factors for risk, spoilage, material costs, etc. as regards all areas of the project.

Thus, in the negotiations, if the quote concerning effort for a particular subsystem is under attack, the project manager could concede the hours quoted for learning and contingency as a trade-off for some other consideration. In like manner, the project manager should know what essential type of hourly effort cannot be reduced without placing the company in a position that would result in an unprofitable contract.

Other areas of negotiation would involve labor rates, overhead rates, and G&A rates, which essentially become a matter of audit. The establishment of labor rates is a matter of record. The overhead and G&A rates are derived by a formula based on statistical and cost accounting data as verified by government auditors or commercial accounting firms.

9.6   Negotiating Specific Contract Clauses

Innumerable variations of contract terms exist. Any procurement organization adopts a group of contract clauses which constitute the procurement policy statements of the organization and are applicable to practically all contracts. These standard contract terms, referred to as contract boiler plate, are generally not negotiable by either party. Since most of the boiler plate terms are commonsense requirements and are characteristic of any business agreement, the offeror would have little reason to try to effect any change in them.

However, the terms of special clauses applicable specifically to the procurement under consideration are another matter, and the obligation that such clauses might place on the offeror can very well determine whether the contract will be profitable to the contractor or whether their presence in the contract imposes such a burden on the contractor that the company would lose money in meeting the contract requirements.

Many of the special contract clauses are intended to clarify which of the contracting parties has the obligation to perform in a particular area. In the case of the RLM simulator, one area of clarification is to establish which party is to be responsible for securing data that would be necessary for simulating the characteristics of the AN/APQ-28 radar. This information must be obtained from the manufacturer of the radar, known to be the Dalar Manufacturing Company of New York, and various government agencies who are concerned with the procurement of the equipment. Since the radar is still in the early stages of production, the data may not be formally documented and probably exists as preliminary drawings, manuscript documents, and other rough forms. Since the effort necessary to obtain the data could be significant, the establishment of the contractual obligation to perform this task is necessary and would be a point of negotiation. If Creative Electronics is to assume this obligation, then the contract should specifically establish this responsibility, and Creative Electronics should negotiate adequate funds and time to carry out this responsibility.

In negotiating a contract, project managers must be acutely aware of the implications that are involved in any special contract clause, and they must establish the minimum acceptable consideration for assuming the contractual responsibility in any area.

9.7 Negotiating Penalty Clauses and Ceiling Contract Price

Penalty clauses relate generally to delivery, whereby a specified amount of money is withheld from the contractor for each day of delay in delivery. A delivery incentive clause is usually associated with a penalty clause, whereby a contractor is rewarded for early delivery on the basis of a specified sum of money for each day that the date of the delivery is bettered.

The project manager must be able to evaluate the penalty and incentive figures against the possible delays that might be experienced or against the improved delivery that might be possible. The actual figures can be readily obtained from those derived in the scheduling plan. The optimistic, most likely, and pessimistic estimates of times calculated for the various PERT (program evaluation review technique) activities could be used as a basis for establishing the range of expected delivery dates and therefore could be used as the schedule limits by the project manager during negotiations.

If, for instance, a penalty figure of $100 per day is being considered for delivery beyond the target delivery date and the most pessimistic delivery derived from the PERT calculations is 60 days, then the project manager must be prepared to consider a $6,000 possible maximum penalty. Then this penalty amount must be evaluated against the incentive, which might be $200 a day for every day that delivery is made ahead of the scheduled date. If the PERT calculations indicate that the most optimistic delivery is 20 days ahead of target date, the maximum possible incentive is $3,000. The project manager must thus strive to negotiate the highest possible incentive rate and the lowest penalty rate.

The target cost is based essentially on the cost estimate as derived in the procedures suggested in Chapter 5. It represents the best effort on the part of the contractor.

In negotiations involving FPI contracts, one very important factor is the amount of the ceiling cost of the contract. The offering of any company such as the Creative Electronics Corporation will include a ceiling figure based essentially on the risk factors inherent in the procurement. The ceiling figure would be a negotiation factor upward or downward, depending on what concessions the project manager must give or receive during the course of the negotiations. If, for instance, the negotiations resulted in a reduction of the estimated engineering hours required and thereby in a reduction of the target cost, then the project manager should counter with a request for a higher ceiling figure for the contract when that phase of the negotiation is reached.

Negotiations involve a large number of variables which the project manager must be able to juggle on the spot. Therefore, the more negotiation tools and guides they possess, the less chance they have of making concessions which might result in an unattractive contract for their company.

9.8 Negotiation Tactics

The first thing to establish at the negotiation table is that the negotiators for the customers have the authority to make commitments for their employers. If this is not established and agreed on, the discussions will merely be an exercise, and project managers should conduct themselves accordingly. In other words, negotiating officials should not entertain the idea of making any concessions if customers will not be in a position to make concessions.

However, assuming that this basic prerequisite is met, the discussions can be initiated. Probably the most important single fundamental rule is to let the other party do the talking to the greatest extent possible. There is a certain amount of psychology in this approach. Generally speaking, the party dominating the conversation of a negotiation meeting is also more prone to concede a contested position. In addition, the party doing the talking will often reveal information of value to the listening party.

There are numerous other fundamentals that should be noted in discussing tactics for negotiating. One is the element of timing as far as making concessions is concerned. Concessions should be held back for the time when maximum benefit might be derived. In addition, points to be made in favor of the offerors or against some requirements which the opponent is attempting to establish should be timed to achieve the greatest favorable impact.

Project managers should camouflage their real objectives to the greatest possible extent if the objective is of vital concern to them. For instance, suppose that the accuracy of the position of the simulated aircraft imposes a very serious design problem for Creative Electronics. Instead of revealing this particular point, the project manager might guide the discussion toward the number of engineering hours estimated in the area in question and be ready to concede a reduction in this area in return for a relaxation of the specified tolerances without making the discussion about accuracy an issue or jeopardizing the procurement.

Since negotiations are in effect a battle of wits, the use of subterfuge is not considered unethical. Because of the general acceptance of this idea, the use of tactics by which one of the negotiating parties might deliberately try to confuse an issue in order to achieve some objective is a common occurrence. Some of the tactics used in confusing the opposition are introducing trivial points to divert attention from some weak or vulnerable point in a proposal, raising a myriad of questions to put the opponent on the defensive, and guiding the opponent's questions to focus on the offeror's strong points.

During the process of negotiation in which each side is striving to gain an advantage at the expense of the other party, pressures and tensions can easily mount and emotional outbursts of one form or another can easily occur. Negotiators who lose control in such a manner can generally be assumed to have weakened their position and, in many cases, to have lost their negotiation objectives. Thus a fundamental requisite of negotiators or project managers is the ability to detach themselves from the issues of the procurement and pursue their aims in a completely objective manner.

As far as the tactics and conduct of a negotiation are concerned, project managers should always keep in mind and recognize that their adversaries have a responsibility to protect the interests of their employers. Since project managers will undoubtedly become involved in future negotiations regardless of the outcome of the one at hand, they must always conduct themselves and their tactics in such a manner so as to command the respect of the opposition.

9.9 Qualities of a Negotiator

Because of the importance of the negotiation phase of a procurement, and because the success of a company in getting a contract award may hinge in large part on how the negotiation is conducted, it is felt that a discussion of the qualities that a negotiator must possess is in order.

A negotiation, by its nature, consists primarily of verbal communication between the two negotiating parties. Thus the first quality that project managers and other negotiating associates must possess is the ability to express themselves and their arguments effectively. There is no implication that these individuals must be silver-tongued orators, but they must possess the facility to translate technical, financial, or other information into clearly understandable language and express their points verbally.

Associated with the ability of expression is the facility to think clearly and rapidly. Upon entering a negotiation, the project manager does not know what stand the procurement negotiator might take or what counterproposal might be made. Therefore, since the negotiation may take a sudden and unexpected turn, the project manager must be able to appraise the implications of the turn of events and how the proposal is affected and then establish a course of action or a counterproposal. This must be accomplished while the negotiations are progressing, with little time for consultation, review of records, or extensive analysis. In other words, members of the negotiating team must have the agility to evaluate the situation and decide a course of action.

Another attribute project managers must possess is the ability to be objective and impersonal in their discussions. The opponent may have a valid point, and project managers should be sufficiently objective to recognize and appreciate the opposing viewpoint. In conjunction with this quality, they must also assume a completely impersonal attachment to their proposal so that any deprecating remarks or criticism will not be taken as a personal criticism.

Patience is a virtue under any circumstances but is particularly important for a negotiator. The ability to subdue one's impulse to speak up when the opposition is attacking one's proposal, especially if the basis of the attack is erroneous, requires a large amount of restraint and patience. However, the patient negotiator should keep in mind that there will be an opportunity for rebuttal and, more important, should realize that in the very act of speaking, the opposition is exposing its position, giving forth valuable information, and in effect weakening its position. Therefore, at every opportunity the project manager should encourage the opposition to speak up and should certainly not try to throttle any arguments being made by the opposition negotiator.

9.10 Summary

Negotiations are conducted to clarify any areas or questions on the technical approaches, cost figures, and other areas related to the procurement. When functioning in the negotiator role, the project manager should strive to counter any objections to his or her proposal, impress the buyer with the merits of his or her offer, and obtain a contract award at the most favorable terms and price for his or her company.

Prior to engaging in a negotiation, the project manager must make a thorough analysis of the technical approaches and cost breakdown of the offering and must be adequately prepared to talk with confidence and conviction on any area. Although the project manager should form a team of individual experts in key areas and have the team participate in the negotiation, the project manager should act as the spokesperson and have complete control over what is said or presented.

Because of the many variables that exist in any procurement, all of which are important, project managers must arm themselves with charts or other readily interpreted references which indicate the boundaries of the variables beyond which they cannot go. The three major variables which must be weighed against each other are price, technical requirements, and delivery schedule.

An evaluation and appreciation of the buyer's position should be based on the best available intelligence in order to serve as a guide for the negotiations.

Of particular importance in any negotiation is the effect that special contract clauses would have on a procurement. The project manager must have sufficient appreciation of the procurement to evaluate the impact of any special contract clause that might be proposed. In general, the project manager should avoid assuming any obligation without adequate contractual consideration.

Negotiation is a complex art in which each party seeks to gain a benefit at the expense of the other. Project managers must be adept at using different tactics for accomplishing their objectives. The nature of the conduct of negotiations demands that negotiating individuals be able to express themselves clearly and logically and be able to comprehend quickly the various complexities that may develop so that they can modify their proposal as required.

10

THE LEGAL WORLD
OF PROJECT MANAGERS

10.1 Contractual and Legal Problems

Project managers and the driver of the family automobile have one thing in common. They both have been given a certain amount of authority, but errors in exercising such authority can result in penalties. The driver of the automobile, though licensed, can be subject to penalties because of damage resulting from negligence. Project managers, by unauthorized direction or failure to act, can violate contractual agreements which would lead to penalties for their organization. Ignorance of the regulations and law in each example will not provide relief from the damages that might be experienced.

In order to avoid legal and contractual complications that may arise from their actions or inactions, project managers must have a working knowledge of contract law and develop a sensitivity toward actions that might lead to contractual complications. In addition, they must know all the special provisions of their contract, the detailed requirements of the specification and schedule, and the type of procurement that is involved so that they can direct their project with the confidence that their actions are legally and contractually correct.

From the project manager's point of view, the rules that regulate the solicitation and proposal evaluations prior to the award of a contract are different from that of a procurement involving commercial parties. The federal, state, city, or other government agencies take the position that the public funds that are used for a procurement belong to every taxpayer—including the companies seeking a contract. Therefore, specific rules and regulations which are designed to provide completely equitable treatment to all qualified companies and ultimately to the contractor govern such procurements.

Procurement actions prior to the execution of legal contracts which involve commercial firms are not subject to the type of regulations that apply to government procurements. The general philosophy is that organizations are free to spend their own money as they wish. Customers are usually not bound by specific regulations when conducting solicitations and may generally award to whomever they wish on whatever basis they desire. In order to discourage unfair practices in solicitations involving private funds, commercial codes often exist which set standards for the conduct of advertised solicitations. However, because of the nature of such codes, companies wishing to participate in such procurements must make their own judgments as to whether they will be treated fairly and whether the contract award will be made on an impartial basis.

Regardless of whether a solicitation for a procurement by the government or private company is involved, it is incumbent on project managers employed by offerors to make a thorough appraisal of their company's chances of success in a procurement prior to deciding whether the money and effort to be spent on a proposal are worth the risks of not being successful in getting an award.

The different environment for the government and the private industry procurement continues prior to and after the signing of the contract. In both cases, the terms of the contract are dominant in establishing the responsibilities of the contracting parties. However, when public funds are used, special rules and regulations of the government agencies (federal, state, city, etc.) place far greater restrictions on what terms can be put into a contract and what either party, particularly the procuring agency, can do.

When contractual problems develop, all attempts to seek a resolution should be made by the two parties. The personnel of the contracting parties have the major responsibilities of protecting the interests of their organizations and at the same time of seeking an equitable solution. If a resolution cannot be reached on a government procurement, the contracting officer can unilaterally direct the contractor to proceed on a particular course. Contractors must proceed as directed but then have the option to formally appeal the direction and prepare a claim against the procuring activity for directing that they pursue a course which is not in accordance with the terms of the contract.

In pursuing a claim, the contractor's argument would be presented to government agencies who sit in judgment of the contractor's claim. Where the contractor has exhausted the channels of appeal provided by the government agencies, the case can be appealed in civil courts. Judgment will then be based upon the same laws and criteria that would be used for contract disputes between private parties.

For contracts involving private parties, the resolution of contractual problems may be more complex. Often a performance bond is involved which protects the customer from damages due to work interruptions, and the issues are referred to the legal courts for prompt resolution. It should be noted that performance bonds are also required for government procurements in many cases.

Figure 10.1 shows the two parallel paths that would be followed in the execution and appeal proceedings of a government and a private industry procurement. The ultimate judgment in either type of contract rests with the civil courts of law as noted in Figure 10.1. What is inferred is that philosophically, the legal aspects of the regulations that govern the conduct of procurements by government agencies must ultimately be able to be supported by the judgments of civil courts.

Since a contract normally involves a legal agreement between two parties, the contractual pitfalls exist on both sides of the fence. Incorrect actions by the project personnel of either the procuring activity or the contractor can be costly to the respective employers. Practically any facet of a contract or any action taken during the planning and execution of a contract can involve legal and contractual complications. The subject is too broad to cover in detail except to identify the most common problem areas in which project managers become involved.

The discussions in this text relate primarily to the legal and contractual world of government procurements, since government agencies are the customers for most large contracts handled by project managers.

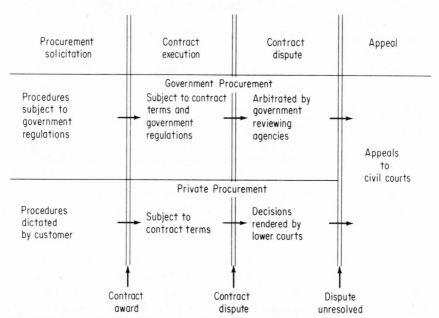

FIG. 10.1 Legal paths for government and commercial contracts.

10.2 Specifications

The writing of the specification and the statement of work (SOW) is usually accomplished under the direction of the customer's project chief. As was discussed, the design specification describes not only what is required but how the equipment is to be designed and built. The performance specification is generally limited to describing what is required so that the details of how the equipment is to be designed and built are left up to the contractor. In either case, the customer's project chief must exercise caution to avoid contractual complications in the writing of the procurement documents. Some of the major specification flaws that might lead to difficulties are as follows:

1. *Omissions:* Whatever is required should be fully described. If a specification contains a void, the contractor could provide a component, system, or design that may not represent what the customer desired; but because of the specification omission, the customer may have no option but to accept what the contractor provides.

2. *Nebulous requirements:* A specification requirement that can be interpreted in more than one way could result in having the contractor provide for the more simple and less costly requirement—generally to the detriment of the customer.

3. *Unclear tolerances:* Accuracies or tolerances are usually expressed in percentages. Disputes can arise if the base for the percentages is not clearly cited. It is essential that tolerances be clearly defined for the entire ranges of operation and that care be taken so that misinterpretations cannot be made from the terms of the specification.

4. *Inconsistencies:* The details of a specification must be consistent throughout. If in one part of the specification a requirement for a performance parameter is less stringent than the same requirement expressed in another part of the specification, the contractor will normally provide the less demanding requirement. Often, what the contractor indicates will be provided will not be satisfactory to the user, but contractually, the customer has no basis for demanding what was needed. The result is that a contract amendment, involving additional funds, has to be negotiated in order to accommodate the customer's requirement.

5. *Impossible requirements:* A requirement may be expressed in the specification or the schedule which, as the project progresses, proves to be impossible to satisfy. A whole battery of legal questions will surface in any such situation, but the one dominant question raised will relate to determining who was the more knowledgeable party at the time of contract. In the absence of any other overriding circumstances, the customer is usually considered to be the more knowledgeable and therefore it has been held that in such circumstances the contractor could not be held responsible for not meeting that particular specification requirement. The point to be made is that if a project manager suspects

a requirement to be impossible to achieve, the information should be brought to the surface for discussion and resolution prior to contract award.

6. *Deficient compliance:* Contractors may deliver equipment which falls short of meeting all the requirements of the specification. The courts have taken the position that if contractors are in substantial compliance with the specification or delivery requirements, the contract cannot be defaulted. However, even though contractors are not held in default, they are not relieved from submitting to an adjustment in contract price to compensate the customer for receiving less than what was contracted. It is incumbent on the contractor's project chief to make sure that all requirements of the contract are satisfied, since the above-noted adjustments could very easily result in a new contract price which represents a loss to the company.

10.3 Proposals and Evaluations

The proposal submitted for a project is the responsibility of the offeror's project manager, and the evaluation of proposals that are received is the responsibility of the customer's project chief. In order to ensure maximum effectiveness of the proposal and realize the greatest economy of time and money, the proposal should avoid addressing anything but that which is identified in the TPR. Since offerors are in competition with each other for the contract award, a proposal should avoid unnecessary embellishments that would inflate the proposed cost. In other words, the proposed design should provide no more than that which is called for in the specification and contract schedule.

The preparation of a technical proposal for a major procurement can be a costly effort for each offeror. Because of cost, most companies may decline to participate in a procurement unless their analysis of the procurement leads the company to conclude that its chance for award is worth the cost involved. The participation decision was discussed in an earlier part of the text. Because of the expense of creating proposals, the government and, in many instances, the larger private customers will purchase technical proposals from qualified offerors, thereby ensuring adequate participation.

The offeror should structure the proposal to satisfy the criteria established by the user. Offering a design that may exceed minimum requirements expressed in the specification is advantageous only if offerors do not penalize themselves because of higher cost. Contract law for government procurements generally establishes that the offeror satisfying the minimum requirements of a specification and having the lowest price would be awarded the contract. If an offeror proposes a unique or improved capability that the customer determines is desirable even though more costly, the customer, particularly if it is the government, will have to amend the solicitation to permit all offerors the opportunity of proposing on the improved capability.

If, with its own resources, a company developed a unique design prior to the date of solicitation of the procurement in question, the company can claim proprietary rights. In cases in which the product of a prior development is patentable, the company has privileges of selling the product to a government agency or private party on an exclusive basis. If the unique design is not patentable, the company can claim proprietary rights on government procurements and enjoy a position of a sole source contractor. When dealing with a private customer, the company would handle the product design as a trade secret and thereby enjoy a position as the exclusive source for the product.

In executing a contract, companies frequently create a design which is unique and could be patentable or at least be handled as a trade secret. Because the development was the result of effort funded by the government, the company would in such a case enjoy exclusive rights only for procurements from other private organizations. For government procurements the development is considered property of the government and the design information is made available to all offerors.

Clarification discussions are often conducted as part of the procurement process. There are two basic types of clarifications: those questions raised by offerors relating to the specification or contract schedule prior to submission of the proposals, and those raised by the customer regarding some point in a proposal. One basic procurement requirement is that all offerors be kept apprised of all information disseminated by the customer. If one offeror solicits and receives an answer to some question on a procurement, government contract regulations require that all offerors be made privy to the questions that were raised and the answers that were provided. The same principle applies to data, and other pertinent intelligence.

The technical evaluation of the proposals almost always involves challenges from the unsuccessful offerors. Generally speaking, the courts of law and the government agencies reviewing protests based on the technical evaluations of proposals have avoided involvement where issues of judgment of professionals are involved. The most common issue of this type relates to the technical judgment of engineers. The courts and reviewing authorities in such cases have generally become involved only in questions relating to procedures.

In commercial procurements, the procedures and rationale for evaluation are established at the time of solicitation and vary from case to case. Legal confrontation can arise when the customer violates ground rules that may have been agreed upon among the customer and offerors, since such violations in themselves may have been a breach of understanding. However, such legal cases are rare. The desire of a customer to preserve its good reputation in procurements is probably more of a deterrent to unfair practices than the threat of possible legal actions.

When a procurement involves public funds, the regulations play a dominant role in maintaining proper procurement procedures. The major points with which the project team evaluating proposals for a government procurement must comply are as follows:

1. The proposals must be evaluated in the areas cited in the TPR document.

2. The criteria requirements for equipment performance which the proposed design is to satisfy must be consistent with the requirements cited in the specification.

3. The TPR document must include information relating to the evaluation criteria and the factors that will be considered for making a contract award.

4. All proposals must be evaluated in the same areas and against uniform criteria.

10.4 The Contract Document

A legal contract comprises two basic elements. There must be an agreement or meeting of the minds between the contracting parties and there must be an exchange of consideration. In the case of a procurement, the consideration would be the goods and/or services the contractor agrees to supply in return for money. The language in the various documents that comprise the contract constitutes the agreement which is attested to by the signatures of the parties representing the contractor and the customer. During technical clarifications and negotiations various points of the specification, schedule, and other documents are usually discussed and clarifying language is established. It is essential that the project manager make certain that the clarifications established are reflected as revisions to the documents that will make up the formal contract. Failure to document agreed-upon clarification into the contract prior to its execution can result in complications later on because of the same issues being raised during the course of implementing the contract.

Once a contract is executed by the parties involved, the courts have ultimate jurisdiction of formal agreements regardless of whether the contract is between private organizations or whether public funds of a government agency are involved. It is incumbent upon the project personnel of the parties to know all the details of the contract in order to avoid placing their company or organization in the vulnerable position of breach of contract.

10.5 Execution of the Contract

The scope of contract law is broad and represents a specialized legal field. As noted earlier, project managers rarely have any formal legal training; but since they have to survive in a legal environment, they must be knowledgeable of the major ground rules of contract law that relate to their project and be able to identify possible legal issues for resolution or pursue courses of action as established by organization attorneys.

The following are some legal points that project managers must recognize during the execution of the contract:

1. *Constructive changes:* The terms of the original contract rarely remain unchanged during its life cycle. Discussions between the project officals of the contracting parties will take place. Some of the discussions could relate to points that represent changes to the contract terms. Any discussion regarding a possible change to the contract should address the impacts of cost, delivery, equipment performance, etc., in depth. If a change is desired, such change shall be reflected as a revision of the specification or schedule and treated as a formal contract change.

However, if as a result of any discussion the contractor acts on a project chief's direction, suggestion, or other expression, a constructive change to the contract may have been precipitated. Such constructive changes usually occur when there is no clear understanding between the contracting parties that any change must be supported by an official documented request or direction. The courts have even stretched the point and ruled that even silence on the part of a project chief with regard to some point that was made by the other party implied agreement and therefore became a constructive change.

Constructive changes can and have resulted in significant changes to the cost, delivery, equipment characteristics, etc., of a contract which were not favorable to the project chief charged with the constructive change responsibility. The point is that project personnel must be very explicit in all their discussions with the other contracting party to avoid having any of their actions result in an unintentional change to the contract.

2. *Termination:* Practically all contracts contain clauses pertaining to the circumstances that may cause the termination of a contract prior to its completion and the legal consequences resulting from such action. The principles relating to contract termination for government and private industry contracts are similar in nature.

Customers can usually terminate "for their convenience," but in so doing are liable for all expenditures incurred by the contractor, including profit and termination charges. Obviously, a convenience termination is an expensive luxury for the customer and would be used only in those rare cases in which unforeseen circumstances of need or money dictate the termination action. The role of project personnel in a convenience termination action primarily involves the establishment of the rationale for negotiating the termination settlement which best serves their employer's interests.

A termination by default action occurs when contractors are unable to complete their contractual obligations or when there is substantial evidence that insufficient progress is being made in prosecuting the contract. Generally, termination by default actions are not valid when circumstances beyond the control of the contractor or "Acts of God" are responsible for the situation.

When circumstances within the control of the contractor, such as poor management, cause the contract effort to come to a halt, the justification for termination by default is clear-cut.

However, when default termination is undertaken on the basis of the contention that the contractor's progress in prosecuting the contract is unacceptable, the issue can be controversial. For such action, the project chief for the procuring activity, who is initiating the termination action, must be prepared to prove to the reviewing authorities and court that the contractor is faced with such insurmountable problems that one or more of the significant objectives of the contract cannot be accomplished. Examples of such problems might include one of the following:

a. Technical difficulty: The design approach adopted by the contractor has lead to problems which, in the opinion of the project chief of the procuring party, cannot be solved within a reasonable time. Because of cost considerations, the contractor may not be willing or able to change the design approach. The customer may determine that the only option left is to terminate for default.

b. Financial problems: Because of money problems, the contractor is unable to perform satisfactorily. If no financial relief is available or bankruptcy is possible, termination action would be warranted.

When a convenience termination is executed, the customer is liable for all costs incurred by the contractor but takes title to anything purchased or produced. In a default termination, the contractor is reimbursed for only those items that may have been accepted. Such items often involve only reports, drawings, or other minor items. Contractors not only have to absorb all incurred costs for undelivered items but are liable for the difference in their contract price and the possible higher price that an alternate contractor may charge to provide the equipment or items originally intended. It is the prime responsibility of the contractor's project manager to run the project so that a default termination is avoided. In the final analysis, the project manager must bear the major responsibility for such a disastrous eventuality.

Some other points of a legal nature which may not precipitate termination actions but can adversely affect the relationship of the contracting parties include the following:

3. *Data and/or furnished equipment:* A contract may require that the customer deliver or make available specific parts or equipment to be used for the product to be delivered or data required for the design of the contracted item. It is the responsibility of the customer's project chief to not only have the data or furnished equipment delivered to the contractor on schedule but to make sure that what is required is complete and adequate for use on the contract. If, for instance, engineering drawings that the customer is obliged to furnish are inaccurate, incomplete, or not delivered as scheduled, the contractor has a legit-

imate basis for a claim against the customer. The customer's project chief must be aware of the legal and contractual implication that the customer has assumed and exert all efforts to live up to the customer's side of the contract.

4. *Progress payments:* Most large contracts which require a long period of time to execute provide for progress payments to be made in accordance with some agreed-on performance criteria. One common criterion or milestone is the completion of the engineering effort, which is the point in the project life when an agreed-on percentage of the contract price is paid to the contractor.

It is vital to the project manager to know in detail the content of the progress payment contract terms and to manage the project to make sure that the progress payment milestones are achieved on schedule. The following are the major reasons why progress payment terms are important:

a. Cash flow enables the contractor to meet current expenses, thereby avoiding the necessity of borrowing money at high interest rates.

b. Meeting scheduled milestone dates is evidence that the project is not experiencing problems. A project that falls behind its projected schedule will also exceed its projected cost—which usually must be absorbed by the contractor.

5. *Penalties, liquidated damages, and incentives:* Delivery schedules on contracts are always important. Often a contract contains terms which penalize late deliveries or provides for a reward for deliveries that are earlier than specified. A contract with penalty clauses imposes major responsibilities on the project personnel of both the contractor and the customer. The contractor must exert every effort to live up to the terms of the contract and in particular to make special efforts to avoid situations that would lead to penalties or damages. The customer's project chief must recognize and live up to any obligations that the contract imposes. If, for instance, the customer fails to deliver data or equipment to the contractor as required, the contractor could claim that its failure to satisfy, covered by penalties or liquidated damage clauses, is due to the fault of the customer and therefore the imposing of penalties is not valid. Case histories have confirmed that the position noted is valid and contractors might thus be relieved of their responsibilities in the area concerned.

10.6 Summary

The regulations for a program using public funds (federal, state, city, etc.) for procurements differ from the rules that might apply to a program involving procurements arranged between private organizations. However, protests that are pursued to the ultimate end would eventually fall under the same legal consideration in the courts of law.

The actions of project personnel can contractually commit the organization or activity they represent; therefore it is essential that they know the terms and conditions of the contract and be aware of the legal consequences of any of their actions or lack of action.

CHAPTER

11

PERT AND OTHER
PROJECT MANAGEMENT
TOOLS

11.1   PERT (Program Evaluation Review Technique) Concepts

The greatly increased complexity, size, and amount of development required by modern commercial as well as military systems have made obsolete the traditional managerial methods and controls for estimating, scheduling, and cost control. In seeking more effective means for managing complex procurements, the PERT and other management systems were developed.

The PERT system is a managerial tool for determining at any point in the life of a program precisely what the status of the program is and where the trouble areas lie.

Its greatest value is that it signals management in advance when any difficulties develop in any specific area which will adversely affect the planned program schedule or budget.

The basic concept of PERT is that the program is divided into discrete, detailed, scheduled tasks which are drawn up into an integrated network. All the significant variables of time, resources, and technical performance are allocated to each task or activity. A system of systematic reporting is then implemented, which enables management to compare actual performance with the original program plan, thereby permitting a continuous check on the program status.

As a management tool, PERT enables the project manager in the line organization or the functional manager in the matrix to shift resources from noncritical to critical activities, thereby permitting the concentration of resources in those areas that were signaled as experiencing difficulties.

11.2 PERT Definitions

The PERT system uses a unique language. The following are the most fundamental terms used:

1. *Activity:* An element of work effort in a program.

2. *Event:* A specific point in the program, usually representing the start or completion of an activity. An event does not have any dimension in time or effort.

3. *Network:* A graphic representation of a program consisting of activities and events which are shown as interconnected paths.

4. *Most likely time, m:* The most realistic estimate of time that it would take to complete an activity.

5. *Optimistic time, a:* The shortest period of time that the completion of an activity would consume.

6. *Pessimistic time, b:* The longest period of time that the completion of an activity would consume.

7. *Expected time, T_e:* The period of time that is predicted for completing an activity. The expected time is statistically derived from the most likely, optimistic, and pessimistic times as expressed in the formula

$$T_e = \frac{a + 4m + b}{6}$$

8. *Cumulative expected time, T_E:* The earliest date that can be anticipated for the completion of a specified work effort or efforts. T_E is the summation of all the expended time T_e along a particular path.

9. *Latest allowable date, T_L:* The latest date on which an event can occur without delaying the completion of the program. The latest allowable time is calculated by subtracting the expected elapsed periods or expected times (T_e) of activities from the date of the last event. If the end date coincides with the date represented by T_E, the $T_L = T_E$.

10. *Positive slack time:* The amount of excess time predicted for the achievement of a particular event. Negative slack indicates the amount of slippage that exists prior to reaching a particular event. Slack time is the difference between the latest allowable date and the expected date ($T_L - T_e$).

11. *Critical path:* The path of a network that requires the longest period of time to complete. This is the path that possesses the smallest positive slack or the greatest negative slack.

12. *PERT work package:* The effort necessary for a particular product, such as a subsystem, computer program routine, document, or other job. The PERT work package is correlated to the necessary activities required for its completion and is identified by a charge number for cost monitoring.

The PERT plan of any major item of a program, such as the hardware and the software, requires that the effort be divided into subtasks, each of which comprises a definable area of work effort. Each of the subdivisions is identified as a PERT work package which constitutes the effort required to complete a specific job and would be represented as one or more activities on the network. It should be noted that the PERT work package is not designed to be identical to the work package created during the operation of the matrix organization, although by coincidence the effort covered in both types of work package may be similar.

11.3   Operation of PERT

The operation of PERT can be divided into the following five broad categories: (1) establishment of objectives, (2) creation of plans, (3) establishment of schedules, (4) evaluation of performance, and (5) arrival at decision and action.

These categories comprise the PERT cycle which is illustrated in Figure 11.1. In addition to the logical flow of sequential information from one category to the next, the corrective feedback loop which constitutes one of the most valuable characteristics of PERT is shown. The corrective feedback permits the project manager to implement changes in the program or plans of action on schedules if the program objective of schedule or cost is in danger of not being met. In addition to being an essential function in setting up a PERT plan for a program, the establishment of the objectives enables the project manager to crystallize the project goal and document the project goals for management and other interested parties.

The RLM simulator program involves several deliverable items in addition to the actual hardware. The other items indicated on the contract schedule are the reports, drawings, and manuals, and a comprehensive network for the program would take cognizance of all these deliverable items. Since this text will

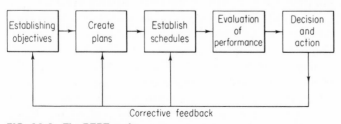

Corrective feedback

FIG. 11.1   The PERT cycle.

discuss only the principles of PERT, in the interest of simplicity the development of the network will be restricted to the hardware.

The creation of the plans involves the translation of the work packages into activities and events which are graphically described as a network. In creating the network, cognizance must be given to the sequence in which each activity is performed and to the earlier events which must be reached before any particular activity starts.

When the activities are set forth, the expected time T_e is derived from estimates made for the optimistic, pessimistic, and most likely times. The project manager would obtain these figures from the individuals who would be responsible for performing the work effort of the activity. The time estimate would be based on using available personnel and resources based on a 40-hour week.

The establishment of the schedule is the conversion of the network elapsed activity times into calendar dates. In deriving the schedule, the project manager must take cognizance of the following basic factors:

1. The contract delivery date and the date on which work is to be initiated.

2. Available personnel and resources of the company (as contrasted with the personnel and resources available for any particular activity).

3. Constraints of different activities. A constraint is an activity that must be completed or an event that must be reached before the work effort of another activity can be initiated. Constraints often present critical areas in a PERT program, since they constitute limitations in facilitating the meeting of schedules.

In drawing up the schedule of the program network, the project manager will probably find that several revisions will have to be made before a PERT plan can be created which will reflect a program planned completion date which is consistent with the contract completion date. The trial and error involved in creating a consistent plan often may involve compromises with what may be the ideal program to the company. For instance, the original plan may have been based on purchasing a particular assembly as a subcontract item having a long lead time. By building the item in-house, the undesirable long lead time might be avoided even if the cost is higher. If the shorter schedule that can be realized by having the item produced in-house justifies the added cost, then the PERT network would be revised.

In the line organization, the evaluation of the performance and the decision and action phases of the PERT cycle is made by each level of management. The officials at each level would study the information derived from PERT from a different point of view and would implement some of the following actions within the framework of their authority: assessment of performance and action, execution of action, and transmitting necessary information about unresolved problems to the next higher level of management as required.

There are numerous types and forms of reports that have evolved from PERT programs. In general, the reports serve to advise management of the program schedule and cost status as of a specific date, a comparison of the actual program status with the program as planned, prediction of program schedule and costs, the identity of areas which are potential or actual sources of difficulty, and other pertinent information of this nature.

In the matrix organization, the management of the PERT is often under the joint cognizance of the project and functional managers.

11.4 Implementing the PERT Plan

The project manager (and function manager in the matrix), in setting up the plan for the RLM simulator project, verify the work breakdown structure. The PERT can be based on practically any desired tier of design detail. The degree of detail of a network is proportional to its complexity, so there is a maximum amount of detail in a network beyond which the amount of the monitoring effort is not proportional to the value of the information derived.

The selection of the events is done in consultation with the individuals of the program team who are to be responsible for the completion of the various work packages. The events must represent specific beginnings or endings of effort in the program.

The PERT network is organized with each event numbered and connected to another event, as shown in Figure 11.2. The arrows indicate the flow of work in a logical sequence. The solid arrows represent actual effort, requiring the completion times shown by the groups of numbers associated with each arrow. The dotted arrows generally represent constraints representing zero time. For

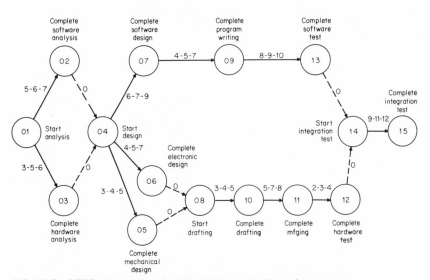

FIG. 11.2 PERT network for characteristics controller subsystem.

example, event 04, start design, cannot be started until event 02, complete software analysis, is completed, even though activity 01-03 could be completed in a shorter time than activity 01-02, as shown in Figure 11.2.

One of the PERT guideline rules is that each activity should be identified with a predecessor and successor event in order to aid in clarifying the network. Since rigorous compliance with the ground rule could result in an excessive and unnecessary number of dummy activities, which complicates the network, judgment in applying the rule is in order. For example, activity 04-05 in Figure 11.2 relates to the mechanical design and is bounded by predecessor event 04, start design, and successor event 05, complete mechanical design.

If the guidelines for predecessor and sucessor events were rigorously applied, the portion of the network which describes the mechanical, electronic, and software design would appear as noted in Figure 11.3, which overcomplicates the network with unnecessary zero activities and events for starting software electronic and mechanical design.

After the project manager has provided for the design of the program PERT network, the next task is to establish the elapsed times required for completing each activity. The sources of this information are generally the group leaders, who will be responsible for the different activities. The three estimates for each

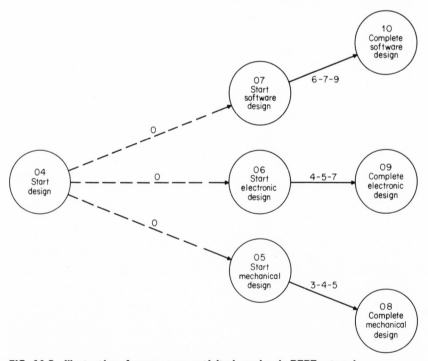

FIG. 11.3   Illustration of excess zero-activity branches in PERT network.

**TABLE 11.1 Summary of Activity Elapsed Times
for Reflectivity Module of Radar Characteristics
Controller Subsystems**

Activity	Identity	Elapsed times	T_e
01-02	Software analysis	5-6-7	6.0°
01-03	Hardware analysis	3-5-6	4.8
02-04	Dummy line	0	0°
03-04	Dummy line	0	0
04-05	Mechanical design	3-4-5	4.0
04-06	Electronic design	4-5-7	5.2
04-07	Software design	6-7-9	7.2°
05-08	Dummy line	0	0
06-08	Dummy line	0	0
07-09	Coding	4-5-7	5.2°
08-010	Drafting	3-4-5	4.0
09-013	Software test	8-9-10	9.0°
10-11	Hardware manufacturer	5-7-8	6.8
11-12	Hardware test	2-3-4	3.0
12-14	Dummy line	0	0
13-14	Dummy line	0	0°
14-15	Hardware/software integration test	9-11-12	10.8°
	Total elapsed time	(critical path)	38.2

°Critical path

activity are the optimistic, most likely, and pessimistic times. These figures are usually indicated on the PERT network as illustrated in Figure 11.2. Activity 01-02, for instance, has elapsed times of 5, 6, and 7 weeks representing the three estimates.

The PERT network is based to a large extent on a statistical analysis. The three estimates for each activity are averaged, with extra weight given to the most likely time to derive an estimated elapsed time T_e, which is used in scheduling.

Table 11.1 lists the different estimated times for each activity of the network illustrated in Figure 11.2.

The PERT network is intimately correlated with the organization of a program, and specific responsibility for each activity must be established and implemented with adequate controls and lines of communication.

The following are some of the inherent characteristics of a PERT network which would be observed when a PERT network system is being designed:

1. Any particular activity must be completed prior to the occurrence of an event. In like manner, an activity cannot be initiated prior to the establishment of an event.

2. All activity paths must be complete and cannot be duplicated or represent alternatives.

3. Any particular event can only occur once.

4. Only one activity line can connect any two events.

11.5   Use of the PERT Network

Once the PERT network has been designed and the expected elapsed times of each activity have been calculated, the project manager can initiate use of the PERT network as a management tool. The status of each event is of primary concern to management, and special attention is focused on the events. An examination of the network in Figure 11.2 will reveal that all activity paths ultimately lead to the final event, but the total of elapsed time T_e differs for different paths. One path requires the largest total elapsed time, and this is the path that is critical to the PERT network. This critical path and the completion of the events on the critical path receive the most management attention.

For the network in Figure 11.2, the critical path is indicated by the double slashes on each activity of the path. The PERT network gives an excellent overall view of the program and enables the managers in the line and matrix organizations to shift personnel and other resources from slack paths to critical paths in order to render aid in overcoming areas of difficulty, as was suggested previously.

Other points to note in the PERT network are that different events will be reached at different times because of the variations in activity estimated elapsed times. The earliest time T_E in which each event can be reached is based upon

TABLE 11.2   Summary of Event Times and the Critical Path for the Radar Characteristics Controller Module

Event	T_E	T_L	Slack	Critical path
01	0	0	0	X
02	6.0	6.0	0	X
03	4.8	6.0	+1.2	
04	6.0	6.0	0	X
05	10.0	13.6	+3.6	
06	11.2	13.6	+2.4	
07	13.2	13.2	0	X
08	11.2	13.6	+2.4	
09	18.4	18.4	0	X
10	15.2	17.6	+2.4	
11	22.0	24.4	+2.4	
12	25.0	27.4	+2.4	
13	27.4	27.4	0	X
14	27.4	27.4	0	X
15	38.2	38.2	0	X

the expected time of each activity. The latest times T_L that events can be reached without jeopardizing the project schedule are calculated by working backward from the final event along the various paths. The various values of T_E and T_L for all events are shown in Table 11.2.

The slack for each event tabulated in Table 11.2 indicates the excess time available to complete the activities leading to a particular event. The slack time is calculated by subtracting the value of T_L from that of T_E.

Positive or zero slack time indicates that all events are at least expected to be on schedule and that no difficulties are expected. If the slack time for any event is a negative value, then the activities contributing to such negative slack times are in difficulty and corrective action of some type is required.

11.6 Probability Features of PERT

One type of information that the PERT network reveals is the earliest time T_E that any event can be expected to be reached, including the final program event. The values of the various event earliest times are derived from the summation of the activity expected times T_e.

An examination of the activity elapsed times in Table 11.1 indicates that there are different spreads between the optimistic and pessimistic times; these spreads are a reflection of the degree of certainty of completing an activity within a specific time. For instance, if the project manager were 100 percent certain that activity 01-02 would take exactly 6.0 weeks to complete, then the three elapsed times would be 6.0-6.0-6.0. The presence of no spread indicates maximum certainty of completing the activity in 6.0 weeks.

The graphic representation of the chances of completing an activity in any of the elapsed times can be shown as the distribution curve in Figure 11.4. If the spread of times is large, the curve would flatten out, and, conversely, a small spread would result in a narrow curve.

The probability of completing the activity in any time measured along the horizontal t axis is represented by the area under the curve at the particular point of interest. Thus the probability of completing the activity within the optimistic time a is very small, and the probability of completing the activity within the pessimistic time b is very large (about 100 percent). The probability of completing the activity within the most likely time is 50 percent, which is the arbitrary chosen basis for the PERT statistics.

One important characteristic of the distribution curve is its standard deviation, which is a direct function of its spread. The standard deviation (SD) is measured from the left and right of the medium (M) along the horizontal axis. One SD designates points on the horizontal axis which are the boundary for 68 percent of the area under the distribution curve, as illustrated in Figure 11.4. Two SDs designate 95 percent, and three SDs designate 99 percent of the area under the curve.

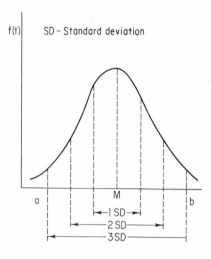

FIG. 11.4 Normal distribution curve typical of PERT activities.

When a series of activities having different time distribution curves is dealt with, the SD of the curves can be correlated. Therefore a value identified as a variance is derived from the SD value by the formula

Variance $= (SD)^2$

The preceding discussion relates to some basic statistical concepts of PERT, and a knowledge of these concepts is necessary for determining the probability of reaching any event in the program within a specific time. This is the type of information that the management would very often ask of the project manager.

In Figure 11.2 and Table 11.2, it is noted that the earliest time T_E in which event 09 will be reached is 18.4 weeks. The time T_E is derived by adding the various values of T_e along the longest or critical path leading to the event. Based on the statistics of PERT, the probability of reaching event 09 in 18.4 weeks is 50 percent.

Assume that the project manager was requested to calculate the probability of reaching event 09 in 17.0 weeks. The procedure for deriving such information is as follows:

1. Calculate the spread of each activity leading to event 09 (see Table 11.3).
2. Calculate the variance of each activity (Table 11.3).
3. Calculate the composite SD of all activities (Table 11.3). Composite SD is the square root of total variance.
4. Substitute values in the following formula to derive factor Z:

$$Z = \frac{\text{Scheduled time} - T_E}{\text{Composite SD}}$$

5. Select the probability for the Z factor from Table 11.4. (Table 11.4 is a range probability representative of a standard distribution curve.)

TABLE 11.3 Summary of Calculations for Probability of Reaching Event 09 in 17.0 Weeks

Activity	Pessimistic time (b)	Optimistic time (a)	Spread $(R; b-a)$	Activity $(SD = R/6)$	Variance $(SD)^2$
01-02	7	5	2	$\%$ = '0.33	0.11
02-04	0	0	0	0	0
04-07	9	6	3	$\%$ = 0.50	0.25
07-09	7	4	3	$\%$ = 0.50	0.25
				Total variance = 0.61	
				SD = 0.78	

$$Z = \frac{\text{Scheduled time} - T_E}{\text{SD}} = \frac{17.0 - 18.4}{0.78} = -1.80$$

The calculation to establish the probability of reaching event 09 in 17.0 weeks instead of the scheduled 18.4 weeks is summarized in Table 11.3.

The actual probability obtained from Table 11.4, which is 0.036, means that the chances of reaching event 09 in 17.0 weeks instead of the 18.4 weeks is 26 out of 1000. The same method of calculation can be made for any other schedule time that may be desired.

TABLE 11.4 Normal Probability Distribution

Positive values		Negative values	
Z	Probability	Z	Probability
0.0	0.500	−0.0	0.500
0.1	0.540	−0.1	0.460
0.2	0.579	−0.2	0.421
0.3	0.618	−0.3	0.382
0.4	0.655	−0.4	0.345
0.5	0.692	−0.5	0.309
0.6	0.726	−0.6	0.274
0.7	0.758	−0.7	0.242
0.8	0.788	−0.8	0.212
0.9	0.816	−0.9	0.184
1.0	0.841	−1.0	0.159
1.2	0.885	−1.2	0.115
1.4	0.919	−1.4	0.081
1.6	0.945	−1.6	0.055
1.8	0.964	−1.8	0.036
2.0	0.977	−2.0	0.023
2.2	0.986	−2.2	0.014
2.4	0.992	−2.4	0.008
2.6	0.995	−2.6	0.005
2.8	0.997	−2.8	0.003
3.0	0.999	−3.0	0.001

11.7 PERT Cost Control

The use of the PERT system to control costs of a program can be effective for whatever degree of detail desired. The most convenient breakdown of PERT costs is based on the engineering, manufacturing, programming, and other costs of the work packages. A reporting system which is correlated with the PERT schedule is used whereby a continuing comparison can be made between incurred and estimated costs for each work package (activities and events).

It should be noted that the PERT cost-control system promotes efficient use of resources, with the resultant economies. The resources scheduled for activities of slack paths can be temporarily reassigned, thereby avoiding nonproductivity while waiting for effort in the more critical paths to be completed. The information that is provided by the PERT is particularly suited to the functional manager in the matrix for scheduling of resources.

Comparisons of the incurred versus the scheduled costs reveal to management the financial status of a project and permit timely action to review areas of cost coverages and to effect means to remedy causes of excessive costs.

11.8 PERT Reporting Documents

Various reports designed to apprise management of the status and trends of a project are prepared and distributed. A report such as the Management Summary Report depicts the cost status and projected time required for a project as of a specific date. Specifically, the report reveals the following types of information to management:

1. *Schedule status:* The difference between the planned and the expected schedule reveals the degree of slippage, if any.

2. *Trouble areas:* The areas which pose a threat to either the planned cost or the schedule of a project are trouble areas.

3. *Cost status:* The comparison of the actual costs against the budgeted cost reveals the cost status.

4. *Cost prediction:* The extrapolation of the actual costs reveals whether a cost overrun or underrun will be realized. The report provides information so that management can pinpoint the trouble areas and apply the necessary remedial effort in time to contain, minimize, or eliminate the causes of the expected difficulty in the project.

The personnel loading reports and displays are useful for planning the personnel requirements and resources for a project. These reports indicate weekly personnel requirements and permit management to meet heavily loaded periods by shifting and hiring personnel. Figure 11.5 presents a typical manpower loading display used for PERT systems.

Two other reports which are generally used for the PERT plan are the cost prediction report (Figure 11.6) and the schedule prediction report (Figure 11.7).

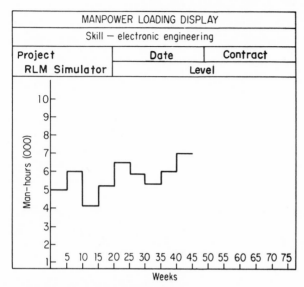

FIG. 11.5   Manpower loading display.

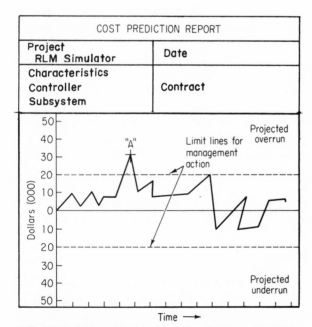

FIG. 11.6   Cost prediction report for the characteristics controller subsystem.

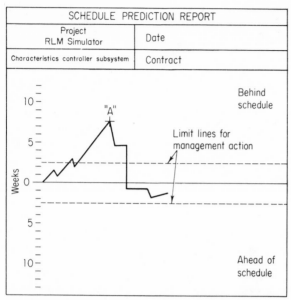

FIG. 11.7   Schedule prediction report for the character-istics controller subsystem.

Because of the close correlation between these two reports, they will be treated simultaneously. On these report forms, the cost and schedule information is plotted for each period in such a manner as to facilitate a projection of the trend of the cost and scheduling of a project.

To illustrate the type of information derived from these curves, a hypothetical situation is assumed in which work on the project has proceeded to a particular point. The normal is represented by the zero axis, and both curves should center around the axis. The dotted lines represent limits or tolerances, and if either curve penetrates the limit line (especially the line representing cost overrun or behind schedule progress), management action is mandatory.

The curves will usually indicate trends so that corrective action can be taken before the curve penetrates the limit lines. At point A, the curves indicate that an adverse situation was experienced and that corrective action was taken. Because of the corrective action, the trend of the curves was reversed, and the project subsystem was brought back within the cost and schedule limits.

11.9   Line of Balance (LOB)

The LOB is a management tool akin to PERT except that it presents only the current project status and has no predictive features such as those in PERT. When applied to a a project involving a single unit or a small quantity, the LOB represents a status report, depicting the degree to which each of the disciplines associated with a project meet the schedule objectives that were established. Originally, the LOB technique was designed for and used on production projects involving large numbers of units. The concept was later modified, and variations of the technique are now used for projects involving the design and development of single or small quantities of units such as with the RLM simulator. Since this text discusses the role of the project manager in managing projects involving engineering and programming, the discussion of the LOB technique will center on principles relating to projects typified by the RLM simulator.

The LOB report for any point in time consists of the following graphic elements:

1. *Objective chart:* A plot of the schedule for percent completed effort (or items in the case of production projects) against time

2. *Progress chart:* A representation of percent completed effort (or items) against type of effort (or production plan for production projects)

3. *Plan:* A plot of types of effort (or production items) against time schedule

To illustrate how a LOB report is compiled and used, Figure 11.8 is presented for the characteristics controller subsystem of the RLM simulator. Normally the LOB report would address the total RLM simulator device, but in order to correlate the information with what was previously presented in discussing PERT, only the above-noted subsystem is discussed.

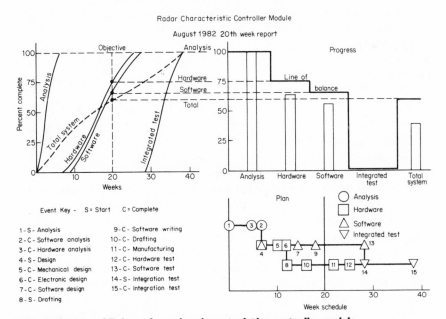

FIG. 11.8   Line of Balance for radar characteristics controller module.

In the objective chart, the schedule curves for analysis, hardware, software, and integrated testing are shown. For the August 1982 report illustrated in Figure 11.8, the vertical 20-week line is shown and the status line representing the percent that should be completed for each type of effort at that point in time is carried over to the progress chart. The LOB for each type of effort is drawn as shown. The shaded vertical columns represent percent actual completion status. The total column reflects the statistical calculation of the percent of completion achieved in the four areas of effort noted in the objective and progress portions of the LOB. The solid line reflects what the scheduled percent of completion should be at the 20-week point. Based on the status of achievement against what was planned (shaded columns versus the solid scheduled line) for each of the areas of effort, the actual completion of the total effort is 37 percent as opposed to the scheduled 57.2 percent. The information revealed by the LOB would be the basis for initiating corrective action by the project manager or cooperative joint actions by the project and functional managers in the matrix organization.

The plan chart portion of Figure 11.8 presents the schedule for various events that are coded in terms of the five categories of effort previously discussed. The identity of the various events was derived from the PERT chart (Figure 11.2) to correlate the principle discussed between the two reporting systems. Normally, if the LOB system is used on a project, the PERT system is not used. The project manager makes the decision as to which system is adequate for the project.

Variations of the LOB reporting system are often used for development programs. However, they all provide the same type of information found in the report presented as Figure 11.8.

11.10 Summary

The PERT system is a management tool designed to control the schedule, cost, and technical performance variables of complex programs and is used primarily where development effort is required. In the traditional organizational complexes, reports to management were historical in nature and therefore precluded timely corrective action on current programs. The PERT system offers a means of comparing current performance and status against planned performance, thereby revealing areas of difficulty and permitting timely corrective action on the causes rather than the symptoms of the problems.

The PERT cycle consists of the elements of determination of objectives, creation of plans, establishment of schedules, evaluation of performance, and the making of decisions. The decision-making effort involves a corrective feedback cycle which results in a very flexible medium for handling unanticipated development and changes.

The PERT work package is the basic building block of the PERT program and consists of a family of activities and events. The degree of control required dictates the amount of detail and the size and number of the work packages.

The network is a flow diagram consisting of activities and events which gives a graphic representation of the requirements of and relationships between various disciplines from the point of view of time and effort. In estimating the length of time required to complete an activity, optimistic, pessimistic, and most likely figures are used.

The critical path of a PERT network reflects the path that requires the longest period of time to traverse. A slack path is one in which time greater than the time required to complete the critical path is available.

The PERT plan is based on the statistical probability that the chance of meeting the schedule with the earliest completion date, for the program itself or any event, is 50-50. The probability of meeting a date other than the earliest completion date can be calculated by applying statistical formulas and using appropriate tables.

The PERT cost control is directly related to the time involved in completing activities required for work packages. Control of cost is accomplished in a manner similar to the way in which schedule control is achieved.

The PERT plan incorporates numerous reporting documents which serve to keep management informed of the program status and thereby to facilitate timely decisions for corrective action.

A LOB reporting system is appropriate for the smaller and less complex projects. It does not have any predictive features but is a useful management tool for providing the status of a project.

There are generally three elements that comprise a LOB report: the objective, progress, and plan charts, which give management the essential information on the project.

BIBLIOGRAPHY

Miller, Robert W.: *Schedule, Cost and Profit Control with PERT*, McGraw-Hill Book Company, New York, 1963.

NAVMAT P1851, Line of Balance Technology, Naval Materiel Command, Dept. of the Navy, Washington, D.C., 1962.

12

INITIATING
THE PROJECT

12.1 Review of Contractual Obligations

The signing of a contract officially executes a legally binding agreement between the buyer and the seller. Project managers must then initiate a contract-monitoring role which will continue until all items are delivered as required by the contract terms.

The framework of the project managers' organization, the scheduling, and the lines of authority have already been established by virtue of their work in preparing the technical proposal and deriving the bid cost figures. The primary and in some respects the sole objective of the project manager is to deliver items which meet the technical requirements within the cost and time schedule of the contract.

As a result of the dialogue that took place during negotiations, the technical requirements for the equipment and the cost and delivery schedule may differ in a number of areas from what was described in the technical proposal. The joint task that confronts the project manager is to review the final documents that make up the signed contract and realign the project plans, resources, and schedule to enable the company to perform in accordance with the official contract.

A specification, particularly if it is a performance type, may have had areas edited during technical clarifications and negotiations. As a result of such changes, adjustments to the proposed price are usually made and clarification revisions are usually incorporated in the proposal. If the technical clarifications necessitate modifications to the proposed design approach, one of the first tasks that the project manager must perform is to make necessary revisions to the scheduled use of resources, the design description, and other established plans that may be affected by the changes.

If for the RLM simulator the negotiation clarification had brought out the fact that the side lobes of the antenna pattern were to be simulated, the project manager would be required to provide for the necessary engineering and programming effort and to seek out adequate resources. The added complexity of that subsystem design might require a specialist and might entail some development work. The project manager may thus have to reassign an engineer possessing the special experience and knowledge required for the performance that was not adequately described in the technical proposal. The same review of the schedule, facilities, and cost budget would be necessary to make sure that the effort that is to be applied on the project will result in producing equipment that meets the contract requirements.

The discussion above emphasizes the importance for project managers to review the final contract details to avoid overlooking requirements that represent contractual obligations. Project managers must also keep in mind that the order of precedence of contractual documents determines any point of conflict. Since the contract schedule and specification take precedence over the technical proposal, the project manager has to recognize that the specifics expressed in the proposal at the time of solicitation would become contractually binding only in situations in which the contract schedule, specification, and referenced documents were silent on any controversial point expressed in the proposal.

12.2 Project Organization

Up to the point of contract award, the role of project manager is one of a staff engineer with temporary delegated authority over other personnel of Creative Electronics. The cost and time consumed in the proposal preparation clarifications, negotiations, etc., are charged as general overhead expenses.

In the matrix organization, the discipline supervisors would make up the basic project team under the cognizance of the project and functional managers and would utilize the resources assigned to them for carrying out the various project tasks as required. The project structure for the RLM simulator in a line organization is shown in Figure 1.2, and the project structure in a matrix organization is illustrated in Figure 1.3.

It should be noted that functions such as contract administration and configuration management are often outside the direct authority of the project manager. One reason for maintaining the independence of those functions is that they both entail responsibilities that extend beyond the domain of any single project, and these responsibilities could be compromised if the functions were assigned to the project managers.

However, if contractual situations evolve which can affect the welfare of the project and parent organization, management decisions relating to the contract could be made which might not be consistent with the views of the project manager, since such decisions would be based on serving the best interests of the parent organization rather than any single project.

12.3 Establishing Tasks and Functions

Before making detailed assignments of work on the project, the project manager will have already completed the updating of the equipment requirements, identified the organization of the project, and established the existence of resources for the job. In the matrix, the resources would be established as a result of actions coordinated with the functional manager.

The identification of tasks usually involves some version of the work breakdown structure which the project manager used in much of the preliminary work on the project. The information in Figure 3.2, for instance, represents a work breakdown structure for the RLM simulator. This was used as the basis for the cost estimate developed in Chapter 5. Elements of the work breakdown structure for the characteristics controller of the simulator were utilized in working out the reporting systems and in illustrating PERT and LOB, as discussed in Chapter 11.

In establishing the tasks for elements of the project, as identified in the work breakdown structure, the project manager is also responsible for establishing the schedule for each task, the budget of man-hours and material costs, the performance criteria, and all other pertinent information that is necessary. To these ends, the information received from the various sources which were used in writing the proposal and preparing the cost estimate would be used to a great degree.

Since the RLM simulator is not the only project being processed by Creative Electronics, the utilization of assignments must be coordinated with a master corporation schedule to prevent conflicts with other projects. In the matrix, difficulties of such a nature are handled by the functional manager. In Chapter 4, however, the overall capacity of the company was presented as a part of the proposal and the analysis indicated that there was ample overall capacity for the simulator program. With this in mind, it will be assumed that the complications presented by conflicts with other projects do not exist. In passing, it should be stated that conflicts in demand for the services of a specialized discipline are a very real problem in industry and have been responsible for large numbers of delays and difficulties that have been experienced on different programs. These situations are particularly prevalent when there is poor overall organizational planning or when slippage occurs on one program which results in conflict with a program scheduled behind it.

12.4 Scheduling of Project Tasks

In the discussion of the matrix and line organizations, it was noted that one of the unique characteristics of the matrix relates to scheduling and the assignment of resources required for a project. Personnel of specific qualifications and other resources, such as production equipment and company computers needed for testing, are scheduled in accordance with the project scheduling plan. Resources for a particular project are allocated by the functional manager as a result of negotiations and discussions with the project manager. The resource allocation procedure of the matrix is designed to achieve full utilization of the company's resources.

In the line organization, personnel resources are assigned to the project team and their efficient utilization is a responsibility of the project manager. Because of the sequence of different types of functions that a project requires, idle manpower is generally unavoidable. In order to control costs, the project manager must plan and coordinate the assignment of personnel resources to ensure maximum utilization of manpower.

Phase charts are one of the planning tools used to provide an overview of the scheduled efforts required for a task or project. Figure 12.1 depicts a phase chart for a characteristics controller subsystem which is presented as a subcontracted item to simplify the illustration. The time frame of the phases of Figure 12.1 is consistent with the PERT network of Figure 11.2 for the module in question. In order to provide as much information as possible, the shaded portions of some of the types of effort shown in Figure 12.1 represent the slack activities that relate to similar activities in the PERT network.

An overall phase chart for the total 130-week schedule of the project would provide a comprehensive picture of interest to management; this chart would be derived from the PERT network of the total RLM simulator system.

12.5 Make-or-buy Decisions

During the precontract phase, the different factors related to whether a component, subassembly, or subsystem should be subcontracted are recognized, and the technical proposal indicates the intent of the offeror at the time of the negotiations (if the TPR calls for such information). Unless the ultimate contract specifies that a subcontract be executed with a particular company for a particular item, the contractor has no legal obligation to enter into a subcontract agreement if events subsequent to the submission of the proposal indicate that the design and fabrication of a particular item could, in fact, be accomplished in-house.

In initiating the project, the project manager must make the final decision as to what items, subassemblies, systems, etc., should be subcontracted. The various factors which previously dictated the decision to subcontract should be reanalyzed, and other questionable areas should be considered. This is particularly necessary if, as often happens, revisions in the procurement requirements have been brought about by the negotiations.

The prime factors which influence the decision regarding subcontracting of a particular article or subsystem include the following: the contractor's capability and experience, the relative cost, the schedule of the contract, loading, future activity, and the customer's desires.

The capability relates primarily to the engineering and manufacturing proficiency of a contractor in a particular area. In the case of the RLM simulator, one area which might be considered for subcontracting is the interface between

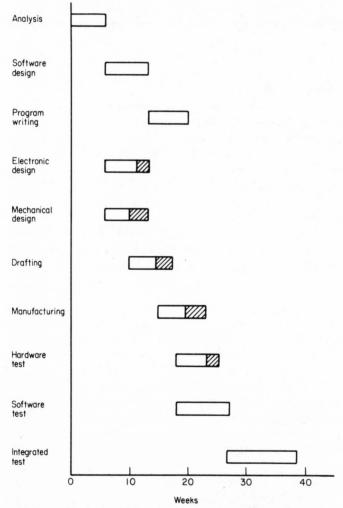

FIG. 12.1 Phase chart for characteristics controller subsystem.

the simulated aircraft and the radar characteristics controller noted in the block diagram of the RLM simulator in Figure 3.3. Since there are numerous signals generated by the simulated aircraft that must be processed by the radar characteristics controller, and since there are many signals originating in the RLM simulator that flow through the controller to the aircraft simulator, the interface is necessary to translate the nature of the signals going each way so as to make them compatible with the system into which they are entering. For instance, the location of the simulated aircraft would normally be expressed in latitude and longitude parameters, and such signals when flowing into the RLM simulator must be expressed in X, Y coordinates of the gaming area of the terrain map. The interface would perform the translation of the interface signals to ensure compatibility with the system into which they enter.

The design and manufacture of a module or subsystem (such as an interface) that requires operational compatibility between two or more different subsystems involve the application of special engineering techniques. The project manager must consider each case to establish whether the required knowledgeable resources are available and whether cost and other considerations would favor a subcontracting or an in-house effort for the equipment under consideration.

Cost is always a prime factor in any decision regarding subcontracting in any area. Everything else being more or less equal, it would be advantageous to perform the work in-house. However, because of experience, know-how, facilities, and other factors, one company may be able to design and/or produce a component or system at less cost than another company.

To demonstrate how the factors of cost would be analyzed by the project manager in making a determination regarding subcontracting, the interface module case will be illustrated. If, for example, the Creative Electronics Corporation has determined that the direct cost and overhead for the module add up to $125,000, the problem is to establish the point at which it would be advantageous to subcontract. Assuming that the direct and overhead costs of an experienced subcontractor for the task add up to $90,000, the following tabulated analysis of the cost to Creative Electronics is presented:

	Produced In-house	Produced by Subcontractor
Direct and overhead cost	$125,000	$ 90,000
Subcontractor G&A, 10%		9,000
Subtotal		99,000
Subcontractor profit 10%		9,900
Subcontractor price		108,900
Creative G&A, 10%	12,500	10,890
Subtotal	137,500	119,790
Creative profit, 10%	13,750	11,979
Total cost	151,250	131,769

With the aid of the preceding table, it can be seen that, in terms of cost of production alone, it would be less costly to subcontract the module.

Another matter that must be considered before reaching a decision regarding subcontracting is the schedule of the program. A comparison of the time required by the prime contractor and the time required by the subcontractor to accomplish a particular task should be made in the same manner as the comparison of costs was made. As was noted in the analysis of costs, increments must be added to the subcontracted effort in order to arrive at a realistic comparison. The schedule increments are time for specification preparation, time for bid responses and subcontract award, time for acceptance testing, and liaison time.

All the increments of time noted above must be added to the time that the subcontractor requires to design and produce a particular item or system. If, for example, the total of the additional elements of time noted above amounted to 8 weeks and the subcontractor's time required to produce an article were 30 weeks, the time cycle for subcontracting would then be 38 weeks. If time were the only factor and the prime contractor could produce an article in 35 weeks, then the obvious decision would be to design and produce the article or system in-house.

Quite often the loading of the prime contractor is such that the personnel and/or facilities are not available to produce an article in a timely manner. If the prime contractor does not wish to invest in additional capacity or hire additional personnel, an article is subcontracted.

An important factor to consider is what future market a particular system or article offers. In spite of undesirable costs, schedule, problems, etc., the prime contractor may desire to produce an article in-house and thereby derive the experience and know-how for future business.

The preference of the customer is another factor that must be considered. A customer may express a preference or even an insistence that a particular item be procured from a particular subcontractor. Whenever possible, the prime contractor should cater to the preference of the customer if the cost, delivery, and performance of the project are not jeopardized. Usually the customer has a sound basis for desiring an article produced by a particular subcontractor and the project would not be adversely affected if such an article were used. If the prime contractor finds that by dealing with a particular subcontractor of the customer's choice the project may be adversely affected, the undesirable effects should be made known to the customer as soon as possible, and an adjustment should be made in the contract cost, schedule, or performance requirements.

12.6 Assigning Tasks

The information in Section 12.4, which dealt with the scheduling of the various tasks of the project, is the basis for the assignment of tasks to the various groups in the project organization. A document identified as a task assignment form is prepared by the project manager and issued to each of the group leaders or discipline supervisors and to the schedule and cost coordinator. There are many variations of the task assignment form, both as to what it is called and as to the form used for the required information. The form illustrated in Table 12.1 calls for basic information that is common to practically all such documents and in this particular case describes the task assigned to the chief engineer of the project.

The task assignment form designates what is required, when it is required, and what the budget to accomplish the task is. There are numerous supplementary documents which provide the necessary detail for the task and also indicate in many cases how the task is to be performed. In addition to the task assignment form, there is a constant communication flow, both written and verbal, to clarify the task requirements.

The chief engineer in turn divides the task into subtasks and makes assignments to various subordinates for execution, giving the necessary information, direction, schedule and budgets. The subdivided tasks not only relate to the various areas of design, as identified in the tier detail shown in Figure 3.4, but also are divided into disciplines of engineering effort. For instance, the radar characteristic controller would involve electronic design effort and mechanical engineering effort for meeting packaging and other requirements.

TABLE 12.1   Sample Task Assignment Form for Project Design Engineer on Radar Landmass Simulator Project

ASSIGNEE: Project Chief Engineer, K. Lawson
TYPE OF TASK: System Design of RLM Simulator Hardware
DATE: 15 January 1983
PROJECT: RLM Simulator, Device 5A1
QUANTITY: One (1) prototype
PROJECT NO.: 6853 Contract No.—N652A
DESCRIPTION OF TASK: Design the Radar Landmass Simulator which conforms to customer's specification (Attachment 1), and Statement of Work (Attachment 3). Design approach shall be consistent with that described in Proposed Design Approach, Radar Landmass Simulator, Project 6853 (Attachment 2).
TASK SCHEDULE: The schedule of the task, to be initiated at once, is to be completed as required by Task Schedule shown in Attachment 4.
BUDGET: Engineering hours—as shown in Attachment 5.
REPORTS: Cost Prediction Report (monthly)
       Line of Balance Report (monthly)
       Design Status Report (semimonthly)
       Manpower Loading Display (monthly)
ATTACHMENT (1): Specification No. 1001
ATTACHMENT (2): Proposed Design Approach, Radar Landmass Simulator, Project 6853
ATTACHMENT (3): Statement of Work
ATTACHMENT (4): Task Schedule, Project 6853, Engineering Design:
RADAR LANDMASS SIMULATOR ATTACHMENT (5): Budget: Engineering Design, Radar Landmass Simulator, Project 6853

The task assignment form for the programming chief contains a different type of task designation than that contained in the form for the project chief engineer but still provides the basic information relating to the schedule, objectives, budget, and other requirements.

The project purchasing agent is responsible for contracting with suppliers for those parts, modules, subsystems, etc., required for the project equipment. Standard parts are routinely handled by issuing purchase orders to those suppliers offering the best buy and delivery to the contractor.

12.7 Summary

The project manager who is responsible for the successful prosecution of a contract has authority over the different areas of required effort. Such authority is depicted in an organizational chart in which authority for the project in a line organization emanates from the project manager.

In the matrix organization, the project manager shares authority over the utilization of resources and their scheduling on the project with the functional manager. Once the allocation of resources required for a project and their scheduling have been worked out with the functional manager, the project manager has the authority to utilize the resources in an efficient manner so as to meet the project objectives.

The meeting of schedule objectives and budgets for the various areas of effort is the responsibility of the project manager. In the line organization, the overall schedule of effort on the project must be phased to minimize both the idleness of persons waiting for some prior work to be completed and the possibility of a jam-up of work due to several tasks reaching a group at one time. As previously noted, the function of the matrix is designed to utilize resources at maximum efficiency.

One particular phase of the project that relates to subcontracting is the weighing of the various factors involved to determine whether a task should be subcontracted or done in-house. The main considerations are the capability, cost, schedule, and capacity of the prime contractor.

Once the organization for the project and the schedule for the accomplishment of various tasks are established, the project manager executes the task assignment forms for various project groups, such as programming and manufacturing; these forms formalize what is to be accomplished, what the completion dates are, what budgets are provided, and in many cases how the task is to be completed.

13

ENGINEERING AND PROGRAM-WRITING EFFORT

13.1 Hardware and Software Relationships

The creation of hardware and software involves many different types of disciplines and procedures, but the result is that both products must ultimately provide for a coordinated, functioning system that satisfies the objectives. Computer software includes the instructions or program for the computer hardware as well as the related documentation, such as the data files, manuals, and other associated supporting material.

For computer systems, the project manager must implement procedures to ensure the following:

1. The meeting of the specification, budget, and schedule objectives of the hardware and software products required in the project

2. The achievement of operational fidelity and harmony when the hardware and computer program are integrated

For the earlier computer systems the programming effort was generally considered a specialized engineering discipline and assigned to a group with essentially no direction other than to complete the task within a scheduled period. Frequently, the computer program failed to function with the hardware in an acceptable manner when the two products were integrated. The troubleshooting of a computer program that is required when malfunctions occur is in itself a time-consuming effort which can be made more difficult if software documentation is not adequately complete and detailed. Projects which experience programming and documentation deficiencies during the final integration phase suffer significant cost overruns, schedule extensions, and failure to achieve performance objectives.

To improve the essential accuracy completeness, fidelity, and other quality properties of the computer software and to provide for greater assurance that the cost and schedule objectives can be achieved, special procedures for designing, program writing, and verifying and testing the software during its creation have evolved. However, even the production of the highest quality software when not correlated with the hardware will result in problems that will plague the success of the project when the integration tests take place. In their efforts to achieve the harmonious functioning of the software and hardware, project managers have sought procedures to verify that the two products are compatible at various points in their development. The different characteristics relating to the desired quality of the design of the hardware and software that are of concern to the project manager are the subject of this chapter. The monitoring techniques and procedures that would be implemented by the project manager to make sure to the greatest degree possible that the quality of the hardware and software and the integrated performance objectives are achieved will be presented in Chapter 14.

13.2 Design Planning

The basic design planning for the RLM simulator equipment was established during the system engineering analysis effort that was presented in Chapter 3. Two basic products of the analysis effort are block diagrams and work breakdown structures (WBS). Since they were prepared for use in the technical proposal, depth of detail was not necessary and the information served to provide only a general overall view of the design approach and the type of effort that would be necessary for the proposed system as dictated by the technical proposal requirements (TPR).

To implement the design effort subsequent to the contract award, a further detailing of the diagrams and WBS is required for the development of the system under contract. The in-depth analysis includes identification and quantification of the various functions, modules, interfaces, and other data relating to the design of the various subsystems of the equipment. Ideally, the details provided by both the block diagrams and the WBS that identify the functional elements of the equipment and the type and amount of required resources have a direct correlation. The work packages resulting from the refining of the WBS represent the level of detail which the project manager requires and uses as criteria in evaluating whether the performance objectives of the project can be achieved.

The planning for the design of a computerized system requires that the project manager give special consideration to development of the software; this is a dimension of production that is not necessary with systems that do not involve computer software design. Although the planning for the design and creation of the hardware and software involves two distinct types of effort by different groups, the project manager is responsible for implementing procedures to maintain close communication between the hardware and software personnel responsible for the design of a particular module or subsystem. This communication is necessary to make sure that the two products are completely compatible.

The specific procedures for assigning engineering and programming tasks to members of the project team vary among organizations, as do the documents used for the assignments. Generally, some form of the task assignment order is issued to each group leader in the line organization or to each discipline supervisor in the matrix organization. A further breakdown of the task assignment order is made which comprises the individual work packages for specific detailed efforts.

The task number will be used throughout the life of the program to identify the project for which work, materials, and parts are expended. The task number is also used for cost control, accounting, and scheduling on the project. The responsibility of the group leader or discipline supervisor is to create a design or program based on the approach developed during the system engineering analysis and described in the technical proposal which will (1) meet the specification requirements, (2) embody the least costly approaches, (3) achieve such characteristics as reliability, quality, and standardization, and (4) be completed within the time schedule.

13.3   Design Data Acquisition

The development and design of a new system requires that data and other types of information be acquired to be used as criteria or performance objectives or to serve in different ways in the engineering process. For a trainer such as the RLM simulator, much of the data would relate to the operational system whose physical and performance characteristics are to be duplicated.

The performance specification for the simulator would identify documents and descriptive material relating to the operational system that is to be simulated by the trainer. Often the data required for a development contract is available primarily through the contracting activity. However, the contractor design engineer is generally the only one who can identify the specific detailed data required for the equipment, and therefore he or she must identify the specific type of data that is required. When, in the case of a government procurement, the government assumes the responsibility of furnishing all data required by the contractor, it assumes an open-ended obligation that can and often does lead to claims by the contractor. On the other hand, the contractor would be at a disadvantage in assuming all responsibility for procuring the data that the design requires, since often specific bits of data are not available to industrial firms for various reasons.

The issue as to what degree of responsibility the government (or other contracting organization) has in assuming the contractual obligation of providing delivery data as GFP (government-furnished property) is one which is constantly debated. Tbe procurement of data for equipment such as a simulator is particularly critical on projects where the data is related to operational systems which are still under development or which have only recently been perfected. Often the data may not be formally documented and therefore is not available from the usual sources. The design of a piece of equipment such as a simulator would require specific data that must be extracted from a mass of general information pertaining to a larger operational system or complex. The contractor's design engineers assigned to the project would be the individuals most qualified to identify the specific data required. Therefore, even though contractors may not be obligated to participate in procuring data, terms are often included in the contract which require that the contractor identify the specific data that is needed and when it must be received.

If for any reason the procuring activity fails to provide the data in accordance with the contract schedule and if the contractor can substantiate the out-of-scope effort that was necessary in executing the contract because of delays in the receipt of data, then a justified claim for funds and/or an extension of delivery can be made by the contractor. In the case of such a claim, the procuring activity generally requires that the contractor cite the specific data that was required and not received on time and document how the delay resulted in the adverse affects on the contract.

The tabulation of the required data is accomplished by the individuals assigned to design the various subsystems. The tabulation is transmitted to the contractor's project manager, who in turn formally conveys the information to the customer's project chief. If it turns out that particular information is not available in its official state, special arrangements and procedures can be implemented by the contractor and procuring activity to jointly obtain the information and data by discussing the requirements personally with a third party—for example, the design engineers of the company building the AN/APQ-28 Radar System. The whole point is that the cooperation and best efforts of both the customer and the contractor are essential in overcoming problem areas (such as data acquisition) and contributing to the success of the program.

13.4 System Design

The creation of any complex equipment requires that the different elements of the system be identified, their functions be established, and the signal flow among elements be determined. The system engineering analysis presented in Chapter 3 constituted the first phase of the design effort and established the basic design approach for the RLM simulator, as described in the technical proposal for the procurement. Because of cost considerations, the depth of the analysis was restricted to the depth and degree of detail that the TPR (technical proposal requirements) document demanded for the offerors' technical proposals. However, the generalized block diagrams, WBS (work breakdown structure), and other technical data had to be valid and accurate, since, as the ultimate successful offeror, Creative Electronics was contractually committed to provide equipment of the design that was described in the technical proposal and that was agreed on by the contracting parties.

The second phase of the system design effort occurs subsequent to the contract award and explains in more detail the basic design approach effected in the technical proposal that was discussed in Chapter 12. In this phase, the primary concerns of the project and functional managers in the line and matrix organizations include the following:

1. Reaffirm equipment performance objectives. Basically, this effort involves recognizing and revising objectives introduced during negotiations and providing a confirmation of previous work.

2. Expand and detail basic design approach. The objective of the effort is to establish the technical data, scheduling, resource allocation, and other data required for documents such as the task assignment orders, the matrix work packages, the engineering orders, and other forms for assigning project work functions to engineers, programmers, and other project team members.

The third phase of the system design effort includes the detailed engineering calculations, program writing, coding, module testing, and verification of the compatibility of hardware and software functions. Whereas the effort in the first phase was essentially system engineering, a transition to detailed engineering and computer programming takes place through the second and into the third phase.

13.5 Hardware Design

Prior to the application of digital computers, the design effort was restricted to engineering the electronic, electrical, mechanical, etc., components of the system. Organizations followed similar procedures for translating specification requirements into design criteria, establishing equations derived from the design criteria, creating functional block diagrams for the system, mechanizing the elements of the system by designing or selecting various modules, integrating and testing the system after fabrication, and performing final system tests.

Design engineers and management echelons had to be knowledgeable of the latest technological advances that could be utilized for the system being designed, but the use of the latest innovations, such as solid-state and integrated circuit developments, did not usually involve any radical departure in engineering procedures. However, it was found that the traditional engineering management procedures were not suitable for the creation of computer software. The project manager is therefore responsible for supervising and coordinating what constitutes separate procedures for the two categories of computer system products.

The application of different types of integrated circuits to computer designs has made possible equipment performance capabilities that could not be achieved in the recent past. Most newly developed systems require the use of digital computers, which must operate with a complete program. Thus designers of hardware for a system must coordinate their efforts with the individuals who develop the computer software. Figure 3.1 shows the software and hardware design to be a coordinated effort established by system engineering; this effort requires integration to provide an operable system.

A computer system such as the RLM simulator is made up of different types of hardware components, some of which must be designed and fabricated by Creative Electronics, and other commercial items, such as general purpose computers that would be purchased from other manufacturers.

The components that would be engineered by Creative Electronics include interfaces: the student station special subsystems, such as the radar characteristics controller; and other subsystems whose function and design is unique to the RLM simulator.

Since one of the primary functions of the hardware in a computer system is to translate the digital signals into some other signal format (to activate other signal-processing equipment or to provide for visual effects, motion, and other physical types of results), the hardware designer must coordinate the engineering effort with the software designer to provide equipment components that will be properly activated by the computer signals.

The details contained in the documents (such as the design work packages, engineering orders, or whatever title an organization may use for the assignment of engineering tasks) would contain sufficient information and reference material to permit the implementation of the task. In addition to carrying out the procedures previously noted for designing a hardware component, the engineer must provide for software compatibility checks of sufficient frequency during the design effort to eliminate or at least minimize the need for design revisions subsequent to the completion of the engineering phase.

13.6 Software Design

The four basic phases constituting the effort required to develop computer programs are depicted in Figure 13.1. Most of the basic analysis information necessary for the development of the computer programs was established during the system engineering phase. However, because of the importance of creating software that is of the highest possible quality, the program manager must make sure that the information and data are complete, accurate, and germane.

Frequently, to accomplish these essential objectives, a programming chief is appointed who has overall responsibility for the software and for providing information to the project manager and carrying out directions.

The programming team headed by the programming chief usually includes a librarian and secretary to maintain the data, test results, etc., and to administer the required communication of the team. The program writers of the team are each given specific assignments to complete, and the programming chief or deputy, if one is included in the team, provides for the coordination and supervision of the programmers' efforts.

In addition to translating the system requirements into computer requirements, the software analysis portion of systems engineering includes verifying the program language to be used, identifying software testing capabilities, developing of test plans, dividing the computer software into modular units for subsequent design work, identifying software configuration items, identifying objectives for the software design specifications, and, in general, establishing the objectives and providing as much data as is feasible to support the software design effort of the next phase.

Among the key products of the analysis phase that are necessary for the initiation of the design phase are the documents relating to the program performance and design requirements. The information in these documents is used to detail the design parameters and to produce the program design specification, which identifies the details of the organization and the design of the computer software.

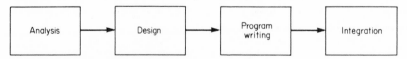

FIG. 13.1   Software development phases.

The types of reports and documents that would be created and used for the software of a particular project depend on the complexity and size of the system, the degree of monitoring required, the contract requirements for documentation, and similar factors. It is the responsibility of the project manager to be aware of what documents are required by the contract, the identity of any additional documents that may be required for the project's effective management (if such reports are not a contractual requirement), and the cost and time required to produce the documents. Table 13.1 lists some of the types of reports that are frequently used in conjunction with the development of computer software.

New or modified techniques for the designing of computer programs are constantly evolving in the search for improved programming methods. The bottom-up approach which was extensively used in the past involved creating individual software modules at the lowest level, performing checks, and proceeding with the integration of other modules as the system was built up. Since the function of the system of modules and the hardware-software integration could not be verified until the top-level programming elements were completed, necessary corrective actions, which were complicated and time-consuming, occurred at the end of the project schedule when time ran out.

The structured software design approach using top-down techniques represents an advance in the art of program writing. The programming hierarchy is

TABLE 13.1   Software Development Documents

Document	Description
Software development plan	Describes the plan and organization to be utilized for the computer program development, design, production, and test effort
Software performance specification (PPS)	Describes performance and functional requirements of software; defines standards and design constraints
Program design specification (PDS)	Describes the design of the software that relates to providing the necessary technical, functional, and performance objectives of the system
Interface design specification (IDS)	Describes the difficult signals that flow among components and the design required to provide compatibility
Program test plan	Identifies the various modules, subsystems, and system tests for the computer program
Program test procedures	Details the specific tests to be used during the development of the computer programs
Data base design document (DBDD)	Identifies type, size, and formats of programming data items; provides memory locations and cross-references to different data items
Program description document (PDD)	Provides functional information concerning the module's time and memory data and other detailed information on the basis of which coding can proceed, with updating as new information is created

started with the top echelon or executive module. Dummy modules are established at the next lower level to facilitate the debugging of the executive module. In an orderly progression, program modules are created at each of the lower levels to replace each of the dummy modules, and sequential debugging tests are made. This structured programming technique provides for the checking of the system as it expands and minimizes the problems related to verifing the complete system at the end of the programming schedule that were often experienced with previous techniques. One significant advantage that the top-down approach offers during the design process is that it is possible to trace the sequence of a particular function from its input through the path of its processing to its output.

Proper use of the software development documents identified in Table 13.1 is essential to the effective development of a computer program. For example, designing involves utilizing the program performance specification (PPS), which is a product of the system engineering effort and reflects each of the functional requirements of the system. The detailed or decomposed information contained in the PPS and other application program design documents, and reflected in the interface design specification (IDS), is translated into the program design specification (PDS). The data in the PDS contains detailed flow charts, descriptions of the method by which each of the broken-down functions will be handled, and other details for implementing the top-down structured program. Another major document in the family of software reports is the program description document (PDD), which provides detailed specifics derived from the PDS to initiate the coding effort and is completed by the coder during the coding effort. The information contained in the various software development documents is summarized in Table 13.1.

One of the more important procedures for monitoring computer software is the design walkthrough. The design walkthrough is usually conducted by independent reviewers, and its purpose is to tell the project manager and other officials whether:

1. The design is consistent with the requirements.
2. The design criteria and philosophy are being met.
3. The design can be implemented and tested and contains no flaws.
4. The quality of the design is acceptable.
5. Adequate test plans and procedures for the design and code are provided.

It is the responsibility of the project manager to keep informed about the latest advances in development of software and to be aware of the advantages of the different techniques of program writing.

Subsequent to the successful completion of the design walkthrough, the critical design review is held with the customer and other interested parties to verify that the program design is capable of meeting the requirements of the performance specification and satisfying any other requirements. The project manager is the key representative of the contractor during the critical design review and is responsible for ensuring the success of the review. The successful completion of the critical design review is the signal for initiating the production or coding phase. The points in the software creation cycle at which the critical design and other reviews for the developing computer program take place are discussed in Chapter 14 and illustrated in Figure 14.5.

The coder translates the functions to be performed by the digital computer into language previously selected to properly activate the computer. The information to be translated or coded exists as mathematical models, flow charts, or similar forms reflecting the required function. The coder translates that information into basic computer commands, such as add and subtract. The information is expressed in the specific languages for which the computer program was designed.

The subject of computer languages is broad and complex. However, the project manager must have a working knowledge of the three basic types of programming language: machine, assembler, and compiler. Of particular importance to the manager is a good appreciation of the different high-level languages, such as FORTRAN and ADA, and the particular characteristics of the language chosen for the computer system for which the project manager is responsible. In addition to taking advantage of the latest programming techniques, the project manager must make sure that the effort includes such features as a thorough analysis, adequate specifications, complete documentation, comprehensive software engineering notebooks, use of software development tools, rigorous testing, and the employment of qualified software designers and programmers and a competent programming chief.

The development of computer software is a complex and demanding effort with many pitfalls. Some basic guidelines that the project manager should follow in carrying out his or her responsibilities are as follows:

1. Establish goals that can be achieved with the resources available.
2. Establish and adhere to a detailed plan.
3. Utilize all available tools and aids to facilitate the effort.
4. Assign responsibility to one individual for the software effort.
5. Proceed in logical steps and correct errors as soon as they are detected.
6. Communicate with all parties who have an interest in the project.

13.7 Packaging Design

With equipment that is not purchased as a completed unit, the mechanical engineer is primarily responsible for meeting the specification requirements for packaging, including the shape, layout, size, and general configuration of the equipment. The mechanical engineer must take into consideration such factors as ventilation of the equipment, accessibility of modules for maintenance, and total weight.

The complexity of modern equipment (especially military equipment) would result in hardware that would be prohibitive in size and weight if design techniques and components used as recently as 15 years ago were to be used today. Because military equipment must be extremely compact in size and light in weight, so-called subminiaturization programs have been undertaken, and techniques using printed circuits, solid state modules, microprocessor circuits, very large scale integrators, etc., have resulted in highly dense designs and compact units. The limitations of size and weight imposed on the simulator are moderate, so that extreme measures for dense packaging are unnecessary. However, the mechanical engineer must complete a design which will be compatible with the specification requirements. If, for instance, the design engineer had planned circuits which could not be incorporated physically into the system and still meet the specification requirements for size and/or weight, the mechanical engineer would indicate the difficulties and the design engineer would have to design the system using a smaller series of modules.

In addition to the responsibilities for providing equipment that meets the physical requirements for size, shape, and weight, the mechanical engineer is responsible for providing equipment that embodies ventilating features adequate for preventing an excessive accumulation of heat generated by the electrical components. The life of components that are subjected to the high ambient temperatures will be significantly shortened. The ability of the equipment to maintain low temperatures is one of the main factors that affects reliability.

Another responsibility of the mechanical engineer is to design the mechanical structure and enclosure so that the specified environmental tests for shock, vibration, moisture, etc., will produce favorable results. The mechanical engineer must not only establish a structural design that will be adequate, he or she must check the major components of the other design areas, such as the electromechanical elements, to verify the ability of the components to meet the environmental tests.

13.8 Drafting

The drafting effort is initiated when the design of a particular subsystem is established and constitutes what is essentially a formalization of the creation of the design engineer. The drawings produced by the drafting department should describe all the detail so that another person qualified to do the work could convert the drawings into operable systems. Because of the amount of detail that must be incorporated in the drawings, a close personal liaison must be established and maintained between the design engineer and those doing the drafting. It would be next to impossible to convey any but the broad description of what is to be incorporated in the drawings via the drafting work order.

Two basic types of drawings are generally required. These are so-called engineering drawings and manufacturing drawings.

The engineering drawings depict the design details of electrical circuits and similar details that are necessary for the maintenance and modification of the equipment after its delivery. These drawings are generally used in conjunction with the maintenance and operation manuals for the support of the equipment in the field.

The manufacturing drawings show all the construction details of the equipment as well as the design details contained in the engineering drawings. These drawings are provided to the manufacturing department for fabrication and assembly of the equipment. Quite often, the customer will use the manufacturing drawings to procure additional identical units of the equipment if the need should develop. In such an eventuality, the drawings are made available to all qualified offerors, and procurement on the basis of a firm fixed-price contract is solicited.

Theoretically, the manufacturing drawings should enable a qualified company to build the equipment with little or no engineering effort. For complex equipment, the idealistic objective is rarely realized, since, when one considers that perhaps a thousand drawings may be involved, it would be statistically almost impossible to have the drawings free from any errors or omissions. Procurements based on such premises invariably result in contractual controversies in which the contractor might submit claims against the buyer because of difficulties and delays encountered due to errors in drawings furnished by the procuring agency. Recognizing this potential source of difficulty, several procurement agencies have protected themselves by incorporating language in the contract which requires all offerors to inspect and analyze the drawings prior to submitting their bid and to provide in the bid price enough of a contingency factor to cover any effort necessary to correct drawing errors and omissions. The contractor thus must provide equipment which will operate as specified in spite of possible drawing deficiencies. This practice has benefited both the buyer and the seller. The buyer has eliminated a cloud of potential claims on the procurement. In a similar manner, any marginally qualified company would hesitate to accept the risk imposed due to the responsibility it must assume for the drawings.

Because drafting is a time-consuming, meticulous effort, close planning and scheduling must be made to minimize the possibility of exceeding the allotted time in the overall project schedule. Therefore, as soon as the general design and the scope of effort are established, the project design engineer will issue to the chief draftsman the drafting order, which will be used as a basis for planning and scheduling. The drafting order includes a description of the type of equipment to be designed; the number of subsystems making up the equipment; the time schedule, showing when each of the subsystem designs will be completed; the estimated number of drawings involved; the list of detail specifications on format; and other pertinent information to which the drawing must conform.

When the engineering design releases are received, the draftsman initiates the work and maintains a close liasion with the design engineer. The information forwarded to the draftsman includes sketches, handwritten notes, and other informal documents which reflect the work and creation of the designer. These documents must be elaborated on and explained to the draftsman so that they can be translated into the formal engineering and manufacturing drawings as required.

In a development program, there are bound to be revisions in the design which must be reflected in the drawings. Any revision that is significant or that may affect the drafting completion date for the particular draftsman must be documented in a drafting order revision.

Drafting is one type of work which does not readily lend itself to any significant degree of automation and which still constitutes a time-consuming, tedious hand operation. There have been some successful attempts to eliminate the necessity of redrawing circuits, structures, or modules which are standard to a company and which are used repeatedly on different types of equipment. An example of such a module would be an amplifier. In such cases, means and techniques have been developed in which the module circuit drawing is available as a transcription which is more or less affixed to the work being processed by the draftsman. Its use eliminates the necessity of drafting a complex schematic or structure. These techniques, many of which are patented, are available, and their use can result in considerable savings in time and cost.

13.9 Value Analysis

A relatively new facet of procurement is called value analysis or, in many instances, value engineering. The objective of value analysis is to provide an incentive to contractors to analyze the product for which they have a contract so as to determine whether a change in design, material, process, or other parameter of the equipment would result in a significant saving in cost without compromising the usefulness of the product. The government and other large organizations that engage in procurement to any degree recognize the human tendency to specify and purchase a particular item over and over again without giving any thought to the possibility that significant savings could be realized by using a substitute. A classic example of value analysis in practice occurred in connection with the procurement by the U.S. Navy of a flexible rope which is snapped between two stanchions at the tops of ladders and passageways. The material traditionally used was a special Manila hemp with a hand-stitched covering of different fabrics. As a result of a value analysis, the substitution of a chain was adopted at a fraction of the cost of the fabric rope.

Although the above example may constitute an overly obvious example of savings through value analysis, it does bring out the point that an open and inquisitive approach to a situation can result in seemingly obvious alterations to existing approaches.

Value analysis of complex equipment is much more subtle and requires a much greater exercise of creative thinking.

Contracts which incorporate a value analysis provision embody a sharing arrangement for any cost savings that may be derived from a successful value analysis. Value analysis is generally used in conjunction with a multitude of production items and has very little application in prototype development units. However, project managers should be cognizant of the potential for profit in a value analysis program so that they will implement such an effort where applicable.

13.10   Summary

The development of a computer system requires that the software under creation be constantly checked and verified to make sure that it functions with the hardware in a compatible manner. The planning of the system design must reflect the interacting functions of both the software and the hardware and must identify the different functions in the block diagrams and the effort required in the WBS.

The acquisition of complete and accurate data is essential for the initiation of the design effort. In addition, the technical approach that was described in the proposal requires review and revision to reflect any changes that were adopted during the negotiations.

The development of computer software invloves completing the four phases of analysis, design, coding, and integration. The analysis is usually completed during the system engineering effort, which involves consideration of both the hardware and the software. Because of the complexity of the software creation, it is essential that the software design be closely monitored and tested as it is developed to ensure its quality and its compatibility with the hardware. To facilitate the software surveillance, various types of documents and reports are generated relating to the design, the testing, and other features of the development process which require review by management officials.

New techniques for facilitating the effective design of computer programs evolve as the use of digital computer systems expands. Modularization of software systems and the use of top-down structured approaches are effective methods for designing, checking, and verifying the computer program as it is developed.

Packaging is a function of the mechanical engineer and involves efficiently arranging components, providing adequate ventilation, and designing a structure and housing that will not only meet the size and weight requirements of the specification but will also permit the passing of the environmental tests.

The chief draftsman is responsible for producing the drawings for the equipment being designed and built. There are generally two types of drawings: engineering and manufacturing. The drawings generally must be created from rough sketches and other informal documents which the design engineer produces. Therefore, a close liaison must be established between the design engineer and the draftsman doing the work so that the finished drawing will accurately reflect the design and construction of the equipment.

Bibliography

Fife, Dennis W.: *Computer Software Management,* U.S. Department of Commerce, Washington, D.C., 1977.

Fischer, K. F., and M. G. Walker: *Digital Systems Technology,* Computer Sciences Corporation, Falls Church, Va., Nov. 1978.

Military Standard Trainers System Software Development, MIL-STD-1644D, Naval Training Equipment Center, Orlando, Fla., 1979.

Summer, C. F.; *Technical Report NAVTRAEQUIPCEN 1089-1 Rev 1,* Naval Training Equipment Center, Orlando, Fla., 1973.

14

CONFIGURATION MANAGEMENT

14.1 Introduction

Configuration management is a procedure designed to accommodate changes required during the development of a system. It was originally applied to hardware systems, but with the increasingly wide application of computer systems, it was recognized that configuration management had to be applied to the development of computer software as well. However, the existing procedures required modification in order to be effective with software because of the following major considerations.

1. The computer program has no physical characteristics, and the requirement for changes and the magnitude of effort necessary for implementing changes are not readily apparent. Therefore, configuration management procedures and disciplines had to provide visibility features, which are essential to the management and monitoring of the programming effort.

2. The changes made in a program module can not only affect other modules, they can necessitate changes to the associated hardware.

The discussion in the first portion of this chapter will present the general principles of configuration management. The latter portion of the chapter contains a discussion of configuration management feature specifically related to computer software.

The two basic types of changes that require consideration during the development of a system are as follows:

1. Changes initiated by external factors, such as revised performance objectives, modified physical requirements, and substitution of improved modules or components

2. Changes required by internal factors, such as correction of errors, revision to accommodate other changes that affect a particular subsystem, and improvements to achieve acceptable performance

In many cases, the implementation of an externally initiated change is determined by an evaluation by the project manager and other officials as to whether the advantages to be realized by the change justify the cost, the schedule adjustment, and other issues that might be involved. However, for mandatory external changes and most of the internally generated changes, no option is available. If a change constitutes on alteration in contract scope, the costs and schedule will require adjustment. If the changes are necessary to rectify errors or to implement improvements so as to satisfy the contract requirements, the project manager must find ways of making these changes that will have the least possible impact on the contract cost and schedule.

In the pre-World War II era, products were less complex and the need to consider changes occurred infrequently. When it did occur, the company was able to get away with an informal system for handling the necessary revisions and modifications. However, for the more complex equipment of that era, the efficient and well-managed companies devised systems for controlling and documenting changes which embraced the underlying principles of configuration management as it is widely practiced today.

During the initial phases of the space program, government authorities soon recognized that loosely controlled changes not only caused complications in their highly complex systems but also raised havoc with other systems and equipment that were designed to operate in close coordination with each other and with the altered systems. Furthermore, since the consideration of changes could be required at any point in time during the life cycle of the equipment, it was recognized that such documents as drawings had to completely and accurately describe the equipment. Many of the complex projects that were pursued in the space programs were experiencing cost and schedule difficulties because the traditional methods for handling changes were not working. Thus the formal configuration management techniques that were then introduced were in large measure responsible for the successful management of the numerous space projects that were pursued and ultimately completed.

14.2 Configuration Management Principles

A broad definition of configuration management is as follows:

> The implementation of formal management and technical direction and controls during a project's life cycle to provide a complete definition of the function and physical characteristics of each item, to control the adoption of changes, and to maintain a continuous accounting of the design and equipment.

Under the configuration management plan, the life cycle of a project is divided into the following four basic phases:

1. Concept formulation phase
2. Definition phase
3. Acquisition phase
   a. Design and development stage
   b. Production stage
4. Operational phase

As shown in Figure 14.1, each phase terminates at the baselines identified as follows:

1. The characteristics baseline terminates the concept formulation phase.
2. The functional baseline terminates the definition phase.
3. The operational baseline terminates the acquisition phase.

In addition, the product subbaseline terminates the design and development stage of the acquisition phase.

There is no identified baseline that terminates the operational phase other

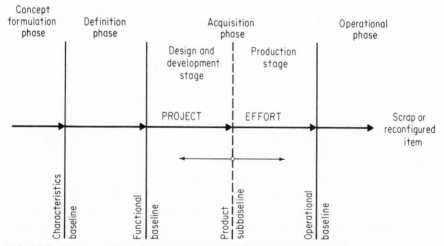

FIG. 14.1 Phases and baselines for configurement management plan.

than to note that at the termination of the operational phase, the equipment is scrapped or reconfigured to the degree where it might be said that the reconfigured equipment assumes a new identity and requires a new configuration management plan.

The baselines represent checkpoints during the life cycle of the project. The initiation of a new phase cannot be made until all details and questions that have been raised during the previous phase have been resolved. For example, if during the concept formulation phase the planners for the RLM simulator were undecided as to whether the size of the simulated area was to be 1,500 by 800 miles or 500 miles square, the next phase, which is the definition phase, could not be initiated since the characteristics baseline had not been established. However, when the point in question is resolved, the documents describing the characteristics baseline would specify the area to be the 1,500 by 800 mile area, and the definition phase could be initiated.

In addition to the configuration management facets of phases and baselines, the disciplines of identity control and accounting must be applied.

The identification discipline requires that during each phase, the complete and accurate description relating to the products of the particular phase be documented. For example, the products of the definition phase would be the specification, schedule, contract, etc., which make up the functional baseline.

Control is the second discipline of configuration management. Proper control ensures that whatever is done during a particular phase is approved by the authorized officials and is accurately and completely reflected in the baseline documents.

The objective of the accounting discipline is to establish an orderly system and procedure for documenting all descriptions and revisions so that complete and accurate data relating to the equipment can be retrieved when desired. Part of the accounting discipline is the audit procedure, which is implemented to make sure that the data describing the equipment is both accurate and complete.

14.3   When Configuration Management Is Required

Practically any company involved in producing equipment has in its system some form of configuration management—particularly if engineering and revisions are normally involved. The degree to which configuration management is applied is often inadequate. Some of the major symptoms of a project which suggest that configuration management disciplines and procedures are not effectively applied are as follows:

1. *Delayed decisions on optional changes:* When the question regarding the merits of a recommended change is raised, a delay in making the decision has an adverse effect regardless of whether the decision is yes or no. Work will be

progressing and if the ultimate decision is to approve the change, the effort to implement the change and the associated costs are usually greater due to the decision delay. The formal procedures contained in a configuration management plan promote promptness in making decisions regarding changes.

2. *Undetected flaws requiring corrective changes:* Inadequate testing and verification of hardware or software that is under development results in failure to identify design or production flaws in a timely manner and compounds the corrective changes that must ultimately be implemented as well as their costs.

3. *Excessive cost of changes:* Changes are often approved without a comprehensive analysis of their cost impact. The procedures inherent in a configuration management plan require an analysis of cost trade-offs.

4. *Documentation that is not consistent with equipment:* The failure to make timely revisions to drawings and documents reflecting the changes that were made to the equipment or computer program usually results in confusion. Time-consuming analysis of the documentation that is not consistent with the equipment is required, particularly when the equipment requires servicing. The procedures provided by configuration management make sure that the revised equipment will be accurately and completely supported by revised documentation.

5. *Objectives that are not clearly stated:* When engineering and programming effort, particularly when related to R&D, is not directed to the requirements of the project, unnecessary and misdirected work can take place. Configuration management defines the specific objectives at the start of a phase and minimizes the possibility of unnecessary effort which does not follow a direct path to the established objectives.

14.4 Concept Formulation Phase

This phase involves the translation of the requirements into descriptive documents, such as the statement of work (SOW) and the function description. It is during the concept formulation phase that the different types of effort are pursued to establish the optimum technical approach, project costs, the time required for implementing the project, the procurement strategy, trade-off studies, and other general characteristics of the project.

In the case of the RLM simulator, the project requirements were established after considering the advantages, costs, and effectiveness of using different training media—for instance, showing students radar videotapes or photographs from actual missions, actually flying training missions, and other means of training. After consideration of the training problems, technical limitations, cost estimates, time frame, and other factors that characterized the various alternatives, the use of electro-optical or digital computer simulator designs to provide the necessary training was determined to best satisfy the training requirements.

If the desired objectives are not feasible because of technical limitations, a research or development program may be pursued to provide a feasible technical approach. The concept formulation phase generally involves the consideration and rejection of ideas, probing, and creative thought. However, the configuration management plan forces the desired objectives to be held in focus by all participants.

Prior to concluding that the concept formulation phase is complete and that the characteristics baseline is established, management should review the results to make sure that:

1. The requirements objectives and performance parameters are defined.
2. The best state-of-the-art technology for the system has been identified.
3. Adequate trade-off analysis had been made.
4. Cost and schedule estimates are realistic.

The concept formulation effort culminates in the establishment of the characteristics baseline, which is represented by some form of requirements document. Until the baseline is reached, no type of effort relating to the subsequent definition phase is initiated. In such a case a configuration management administrator or committee would be charged with the responsibility for making sure that the documentation relating to a piece of equipment is acceptable and meets the criteria for establishing the characteristics baseline. It should be noted that the formulation and definition phases are implemented by the customer and that the acquisition phase is implemented by the contractor. No one individual functions as project manager on all phases except when an organization undertakes all phases of a project as a total effort.

14.5   Definition Phase

The primary objective of the definition phase effort is to translate the requirement that the characteristics baseline establishes into the equipment specification and other planning documents that will satisfy the requirements for the functional baseline, as conveyed in Figure 14.1. The type and detail of the documents created during the definition phase depend on the particular project involved. If the equipment will be procured from a contractor who performs the design, development, programming, fabrication, testing, etc., the functional baseline will be represented by such documents as a performance and contract end-item specification, a contract schedule, software documentation, and proposal requests. If, however, the equipment will be produced as an in-house effort, the baseline will include a design specification, a detailed plan for executing the various operations of the in-house effort, etc.

Any piece of equipment or system that is to be created must be supported by such items as operation manuals, computer programming data, maintenance handbooks, drawings, and spare parts. The number of support items is usually proportional to the complexity of the equipment. A kitchen toaster, for instance, will require only a simple booklet of instructions, whereas a complex computer system usually requires drawings, program listings, flowcharts, training courses, special test equipment, etc. All items of a project must be defined and scheduled during the definition phase. In configuration management, such items are called contract end items (CEIs) and computer program configuration items (CPCIs). They are described in the contract end-item specification. For the RLM simulator project, the CEIs and CPCIs would be designated as the actual line items of the contract.

Some characteristics of these items are as follows:

1. Capable of being described by a specification
2. Identified by top drawings
3. Capable of being documented
4. Capable of being correlated with other end items

The primary objectives of the definition phase are to translate the requirements established during the concept formulation phase into specific documents that describe the equipment to be created; these documents constitute the functional baseline.

If the equipment is to be procured from an outside organization, as in the case of the RLM simulator, the effort for the definition phase includes proposal solicitation, evaluations, negotiations, and a contract award.

Since changes and revisions are often generated and adopted during the negotiation process, it is highly important that a management review be made of the results of the definition phase. The following are some of the functions of such a review.

1. Review of risk factors to establish whether any new factors have evolved
2. Confirmation of the content of the specification and other contractual documents
3. Confirmation of cost and delivery estimates and terms
4. Review of the factors which determined the choice of the contractor

14.6 Acquisition Phase

The word "acquisition" implies that the equipment will be procured under contract. Whereas contracting is the usual means for obtaining equipment, the disciplines involved in the acquisition phase also apply to an in-house effort for designing and producing an item. For the purposes of this discussion, acquisition by contracting will be assumed.

As noted in the configuration management plan in Figure 14.1, the acquisition phase is divided by the product subbaseline into the design and development stage and the production stage. The various documents that represent the functional baseline are used to launch the design and development stage. During this stage of the acquisition phase, the products of the engineering, computer programming, drafting, and other similar types of effort are documented in engineering and manufacturing drawings, programming flowcharts and design specifications, production specifications, and production planning schedules. When approved, such data constitute the product subbaseline. Once the product baseline criteria have been met, the production stage can be initiated.

In most projects, establishment of the product baseline does not occur at a single point in time. The design of hardware subsystems and the programming of the different modules occur at different rates, and the data and materials that in each case may be required make their appearance at different points in the schedule. The utilization of the PERT system assures the project manager that the different efforts are being tracked and controlled. There is a close correlation between the PERT or other management scheduling tool and the baseline dates of the configuration management plan. However, before any subsystem can be released for manufacture or before any program design can be assigned to a coder, the project manager must verify that the product subbaseline for the particular subsystem or software module is acceptable.

The primary effort during the production stage involves the manufacture, programming, fabrication assembly, testing, and acceptance of the equipment required by the project. The operational baseline is established with the CEIs or CPCIs and, in particular, when the equipment under procurement is delivered and accepted and when verification is made that the acquisition phase objectives that were documented at the functional baseline are satisfied.

Testing to verify the design of hardware subsystems and modules of the equipment under development is an integral part of the acquisition phase. In earlier times, such tests were generally performed at the discretion of the individual designers on breadboard circuits and were treated as a part of the designing process. Before the application of computer technology, schedules and procedures for testing hardware were identified in project plans subsequent to the completion of fabrication of the different subsystems and were identified as in-plant system and acceptance testing.

To avoid the many software quality problems of the early programming efforts, comprehensive debugging, verification, and testing procedures are implemented during the design and program-writing phase of a software development effort. Validation testing of the complete software system and the integrated hardware-computer program systems follows. It is the responsibility of the project manager to recognize and plan for the comprehensive testing effort that the creation of software requires. In the matrix, the scheduling of resources for testing is the joint responsibility of the project and functional managers.

Before the operational baseline is established, the project manager is responsible for reviewing the results of the acquisition phase to make sure that the required objectives were achieved. The following are some of the items that must be checked:

1. Completion of all tests
2. Status of software and hardware documents
3. Verification that the equipment meets performance objectives
4. Appraisal of project costs
5. Review of support capabilities
6. Certification of the accuracy and completion of design and program changes

14.7 Operational Phase

The utilization of the equipment that has been produced during the earlier phases occurs during the operational phase. This is the primary test of whether the physical and performance characteristics of the equipment satisfy the objectives that were established and approved at the operational baseline. Another major objective of configuration management is to provide consistency between the equipment and the supporting documentation. If the drawings, manuals, programming documents, and other support items completely and accurately reflect the hardware and software that are being utilized, then the objectives have been satisfactorily met.

The ease with which computer programs can be modified, the fact that software revisions are difficult to locate without a precise record of the change, and the fact that programming revisions that are made to correct a problem in one area may adversely affect functions in other areas of the system make necessary the rigorous control and documentation of all software revisions. The procedures that the configuration management plan provides are particularly important for the operational phase, when the original designers and programmers are not available at the site.

The disciplines applied during the operational phase (as discussed in Section 14.2) continue for the life of the equipment. Changes to the hardware and software must be reflected in the drawings and other types of documentation. The configuration management effort ceases only when the equipment is scrapped or a redesign effort of such scope is applied that the equipment loses its former identity. Under the latter circumstances, an entirely new set of objectives would be established and a new configuration management plan would be adopted.

14.8 Disciplines of Configuration Management

The three disciplines of configuration management, as cited in Section 14.2, are indentification, control, and accounting. During the early phases, identification plays a major role, since it is during the concept formulation phase and to a lesser degree during the definition phase that ideas, concepts, and objectives are formulated. The control and accounting disciplines can only be applied when the concepts, etc., take form and are documented. Therefore the control and accounting disciplines become dominant during the acquisition and operational phases.

Figure 14.2 presents a flow diagram of the identification process that might be used by the project manager for an in-house development or by the project manager for a procuring or using activity to establish technically feasible objectives during the concept formulation phase. The critical effort is the evaluation of technical approaches for a set of requirements. It is at this point that one of the following five alternatives can ultimately be taken. The alternatives are keyed by numbers to the different paths depicted in Figure 14.2 as follows:

Path 1: All the capabilities or requirements are technically feasible and can be documented to establish the characteristics baseline.

Path 2: The existing technology is considered inadequate for meeting the requirements that were established and an R&D effort will be necessary.

Path 3: The results of the R&D effort were successful in achieving a technology that was capable of satisfying the requirements, and therefore the project can continue toward the original requirement objectives.

Path 4: The R&D effort did not result in deriving a feasible technical approach for one or more of the required characteristics. In such a case, the project can continue if there is a high level of confidence that technically feasible approaches for the requirement can be successfully developed during the definition phase. However, the costs of, and risks involved in, successfully developing the technical solution for the requirement that was judged unfeasible due

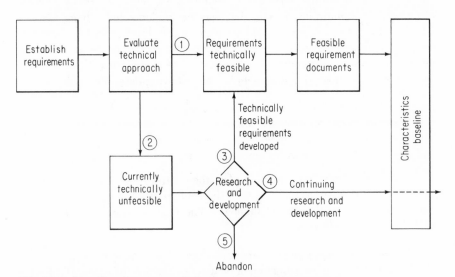

FIG. 14.2 Identification process during concept formulation phase.

to technological limitations increase as the project progresses into the definition phase. An organization must weigh the possibility of expending time and money for a technical solution that may not be achieved against the potential profits that would be realized if the R&D effort were to prove successful. Failure to achieve a breakthrough would probably mean that the project would have to be aborted.

Path 5: The fifth possibility is the judgment that some of the requirements are not feasible and cannot be met within any realistic time frame and that the requirement in question must be abandoned.

The point to be stressed is that by using the characteristics baseline as a checkpoint, the possibility of spending time and money to get results that may not be technically feasible is all but eliminated. In addition, the focus and objectives of R&D effort are identified, and a time frame for achieving such objectives is established.

During the definition phase, the identification discipline becomes more concrete. Identification involves the translation of the documented requirements that have been established as technically feasible into specifications and other planning documents.

For the acquisition phase, identification involves the translation of specification and contract documents into work packages, engineering drawings, and job orders and ultimately into the equipment itself. For the operational phase, the identification discipline is minimal, since it would apply primarily to changes to the equipment which must be reflected in the drawings and other documentation.

The control discipline in configuration management relates primarily to the direction of the effort necessary for achieving each baseline and for evaluating, selecting, and monitoring changes that may be implemented. Changes in the design of the system hardware and software under development are essential for various reasons, including the need to correct errors, improve the design or reliability, revise requirements, etc. The control of changes that the configuration management plan provides includes the following desirable results:

1. Avoids changes that are unnecessary
2. Establishes the priority of changes
3. Ensures accurate documentation of changes
4. Provides timely action for approved changes

In the configuration management plan, the control functions are vested in a special monitoring group referred to as the configuration control board (CCB), which is chaired by the project manager. The primary responsibilities of the CCB include the following:

1. Review and authorize or disapprove release for proposed changes.
2. Ensure adequate coordination of changes between hardware and software systems.

3. Ensure representation of all parties concerned with the project, such as procurement representatives, users, designers, etc.

4. Provide for accurate records of proceedings and decisions.

5. Ensure follow-up of documentation, contractual, scheduling, and other actions.

The accounting discipline is closely related to control and is most heavily used in the acquisition and operational phases. The accounting discipline, often computer-automated, serves to provide the following:

1. Keep documents up to date by recording changes as they occur.

2. Monitor the execution of changes authorized by the CCB.

3. Provide a current record of data in order to provide the capability to retrieve information as required.

4. Implement configuration audits to verify that the documents which describe the equipment are complete and up to date.

The major products of the configuration management disciplines to which status accounting applies include the following:

1. Specifications and revisions

2. Manuals, software documents, test procedures, etc.

3. Status of reports of changes

Reviews and audits are essential configuration management functions for managing a development project. Reviews of project milestones are periodically conducted to assess the progress of the project. The audit function establishes the degree to which the hardware and software designs satisfy the specification and other data requirements.

In essence, the configuration audit is conducted to certify that the configuration and performance of the equipment with its subsystems and of the software are consistent with the baseline documentation. Audits are normally conducted at the various baselines before approving the baselines. Minor audits relate primarily to end items and can be conducted at any point in the life cycle of the project.

For example, the product baseline audit that occurs during the acquisition phase determines the following:

1. Compatibility of the functional and physical characteristics of the equipment or end items with the documents

2. Validity of the equipment testing to verify compliance with the specified requirements

3. Whether the documents (such as drawings) reflect all approved revisions and changes

The operational support baseline audits are conducted to:

1. Establish that the drawings and data completely and accurately reflect the equipment, including approved changes

2. Identify areas needing change and the revisions that are required

The responsibility and authority to conduct audits rest with the project manager. On projects which include configuration management personnel, the project manager frequently delegates the audits to a configuration management official.

14.9 Changes

One of the major objectives of configuration management is to make sure that when changes are made to the equipment or computer program, such changes are accurately reflected in the documents that support that design. The chaos and frustration that result from the situation in which the drawings or other documents do not accurately describe the equipment can be a major cause of unnecessary, time-consuming expenditures.

Firgure 14.3 illustrates the basic procedures involved in making sure that approved changes to the equipment are faithfully reflected in the documents that support the equipment. As noted earlier, the characteristics baseline establishes the requirements for the equipment. It is the objective of the configuration management program to make sure that if any approved changes are implemented during the life cycle of the hardware and software system, the documents of the characteristics baseline will be updated as appropriate to reflect such changes. If, in the case of the RLM simulator, a change were approved during the definition phase which increased the area to be simulated,

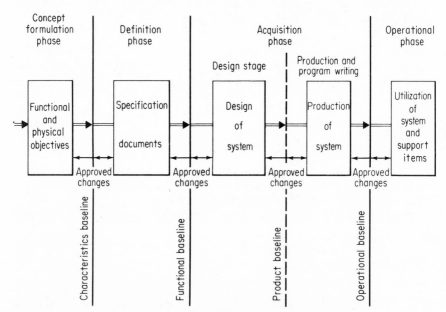

FIG. 14.3 Effects of changes on configuration management baselines.

such a change would be fed back to be reflected in the characteristics documents. The procedure thus makes sure that there will be no inconsistency among the documents that will eventually serve as the functional baseline.

The same procedure would be followed during any of the subsequent phases or stages. In summary, the results of the procedures for handling changes during the configuration management life cycle include the following:

1. During any phase, consistency among the documents and/or system being created during that phase and the previous baseline

2. Consistency among the documents that serve as the various baselines

3. Consistency between the system and support items that are utilized during the operational phase and the documents that serve as the operational baseline

The primary reason for implementing configuration management for a project of an organization is to facilitate the achievement of cost, schedule, and performance objectives. The disciplines of identification, control, and accounting can be tailored to the size of the project. Configuration management can be effectively applied to all sizes of organizations or projects. Some of the benefits that can be realized are:

1. *Establishment of objectives for each phase:* The baselines and the disciplines that are applied to such baselines eliminate the expenditure of effort on incorrect approaches or for unfeasible objectives. Once the baseline has been formally approved, the objectives for the next phase have been documented and the possibility of proceeding down the wrong path has been all but eliminated. The establishment of each of the successive baselines points the way to the next set of objectives.

2. *Effective channeling of resources:* The benefits derived from the channeling of resources are particularly valuable during the concept formulation phase, since the effort relates to abstract notions and requirements. In the other phases the channeling of resources is facilitated, since the establishment of each of the baselines points the way to the objectives of the project.

3. *Elimination of redundant effort:* The clear establishment of objectives and channeling of resources act to eliminate redundant effort. Redundant effort usually involves two forms: work that more than one group is pursuing and work that must be repeated because of an erroneous approach or an erroneous objective. The implementation of the various controls, such as design reviews and audits, acts to detect errors and wrong approaches before they go too far.

4. *Facilitate retrieval of accurate data:* The configuration management change procedures require that only approved changes be made to the equipment and that when they are made, the documents be revised at the time of change. The accounting discipline which provides for the retrieval of equipment data when required serves to eliminate wasted effort and unnecessary expense.

14.10   Software Configuration Management

In the introduction to the subject of configuration management, it was noted that because of the unique physical characteristics of a computer program and the effect that software changes (which can be easily made by personnel having access to the software) have on the function of the computer system, special procedures to implement configuration management are required. The basic principles of configuration management apply for software, but more comprehensive tests, checks, documentation, etc., have to be designed to implement each of the configuration management phases and to achieve the different baseline objectives.

In addition to providing for the verification and validation of the software itself, a constant check must be made during the software design and coding process to ensure compatibility with the hardware design.

Figure 14.4 identifies the basic functions required to create a computer system. The primary objective of the configuration management plan for the project is to achieve system performance that meets the requirements.

Figure 14.5 depicts the points in time in the various phases when those different types of actions—reviews, tests, and audits—would normally be implemented by the project manager. An explanation of the major functions of the configuration management of a computer system noted in Figure 14.4 are as follows:

1. *Software requirements review (SRR):* Review of all the system software and hardware requirements, covering results of system engineering, trade-off studies, support, and other requirements

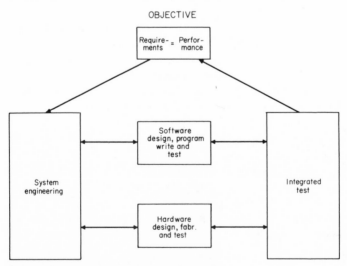

OBJECTIVE

Require- = Perfor-
ments   mance

Software design, program write and test

System engineering

Integrated test

Hardware design, fabr. and test

FIG. 14.4   Computer system creation cycle.

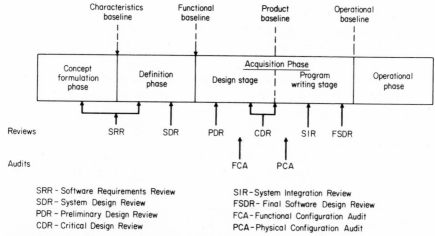

FIG. 14.5   **Configuration management software functions.**

2. *System design review:* Early assessment of the overall computer systems to make sure that the best design for the requirements is adopted, that risks have been identified, and that the system is cost-effective

3. *Preliminary design review (PDR):* Technical review to evaluate design approach, compatibility with program requirements, and plans for testing of software and hardware systems

4. *Critical design review (CDR):* Confirmation of the software design prior to release for coding

5. *System integration review (SIR):* Verification that software and hardware are acceptably integrated

6. *Final software design review (FSDR):* Review of design, associated documents, and acceptance test procedures plans

7. *Functional configuration audit (FCA):* Verification of software module performance in relation to specification requirements

8. *Physical configuration audit (PCA):* Assurance of the correlation between coded software modules and documentation

It is the responsibility of the project manager to adopt a configuration management plan that is tailored to the project and to enforce the procedures and controls. The effective implementation of configuration management plans for both the hardware and software will provide the means of realizing the goals of the project.

14.11   Summary

Configuration management is a procedure by which changes required for the hardware or software portions of a system under development are monitored and documented to ensure complete visibility of the products and consistency between the documentation and the products themselves. Required changes are characterized as external when they relate to requirements and internal when they involve the correction of errors or adoption of improvements.

Configuration management plans, which must be tailored to the project by the manager, comprise the four phases of concept formulation, definition, acquisition, and operation. The acquisition phase is divided into two stages— the design and development stage and the production stage.

Baselines, which separate the different phases and stages, represent points in the configuration management cycle which provide a precise identity of the results of the preceding phase and the objectives of the following phase. The baselines are identified as the characteristics, functional, product, and operational baselines. The effort on a project cannot proceed until the documents constituting each baseline are verified as complete and accurate.

The three disciplines that are applied to varying degrees in each phase and stage are identification, control, and accounting. A computer is generally used to perform the control and accounting functions on larger programs.

Because of the intangible characteristics of software and the ease with which uncontrolled an unauthorized changes can be made, the configuration management plans for the development of computer programs in corporate identification, control, and accounting procedures are more rigorous than plans used for hardware systems. In addition to making sure that the software itself is acceptable, correlation between the software and hardware systems is provided to make sure that the effects of software changes on hardware performance and vice versa are recognized and accommodated.

BIBLIOGRAPHY

Gaylor, W., and R. Murphy: *On-line Configuration Management: Proceedings of the Interservice/Industry Training Equipment Conference*, National Security Industrial Association, 1980.

Military Standard Trainer System: Software Development, MIL-STD-1641, Naval Training Equipment Center, Orlando, Fla., 1979.

PROJECT MONITORING, COMMUNICATION, AND DECISION MAKING

15.1 Communication of Problems

In order to prosecute and complete a project successfully, it is mandatory that an effective system be provided for the communication of relevant information among assigned and interested people. Three of the most vital reasons for establishing an effective communication system are to provide information for decisions, to issue instructions or guidance, and to provide a means of reporting on the project status.

The communications needed to come to a decision usually are initiated by the existence of a problem that must be solved. Figure 15.1 illustrates the dual communication cycles in the management and working-level areas and the various elements making up each cycle which relate to a typical project in the line organization. It should be noted that the one element which is common to both cycles is the project manager's instruction function.

Figure 15.1 shows the internal communication cycles relating to instructions that involve the project manager. The project communication cycle of the working level occurs in the following sequence:

1. The instruction is issued by the project manager to the working level.
2. The party responsible for putting the instruction into effect is represented

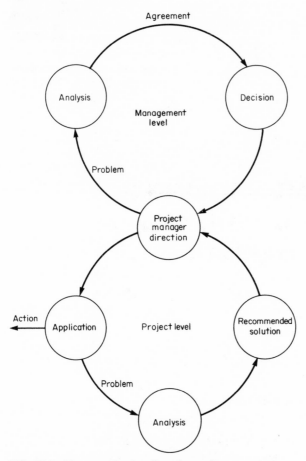

FIG. 15.1 Line organization direction cycles.

by the application element. If the execution of the instruction does not meet with any problem, the action is completed.

3. If, however, the carrying out of the instruction meets with problems, an analysis is necessary. The analysis involves the gathering together of information relative to the problem, which might require inputs from sources not shown on the illustrated communication cycle.

4. After the analysis is completed, the recommended solution is given to the project manager.

5. The solution is translated into an instruction by the project manager and the cycle is repeated. If everything has worked well, the application of the instruction will result in the action.

Occasionally project managers may not be able to translate the decision into an instruction because of complications contributed by factors over which they have no control. For instance, the decision might be to utilize certain facilities of the company at a particular time. It may be that a conflict over use of the facilities exists with another project in the company. In such a case the project manager would refer the problem to the management-level cycle shown in Figure 15.1. The decision rendered in the management-level cycle would be referred to the project manager for conversion into an instruction for the working-level cycle.

In the matrix organization, the project communication cycles are more complex, as noted in Figure 15.2. Since the solutions to problems often affect the resources that are planned for the project and the schedules, revisions concerning resources must be coordinated with the functional manager. Usually agreements are worked out and the instruction is rendered by the project manager as in the line organization.

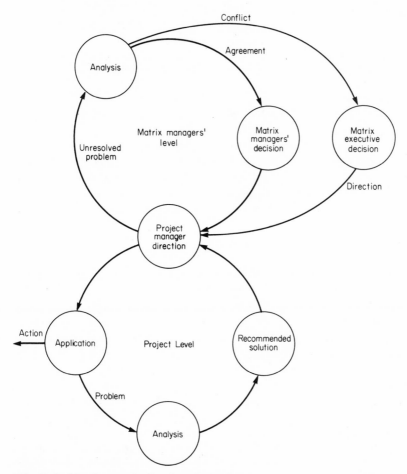

FIG. 15.2   Matrix organization direction cycles.

Even in the most effective matrix systems, conflicts between the project and functional managers require arbitration at the executive levels. Figure 15.2 illustrates an additional communication link that requires direction by the matrix executive and is binding on both the project and the functional managers. In such situations, adjustments to the project objectives are required and the executive direction is translated into instructions issued to the project team by the project manager.

15.2 Project Status Communication

Communications to reveal the status of a program should flow from the working level up to the top echelon of management and down to the lowest levels. The four basic points that the status reports must address are:

1. Are the technical requirements of the specification being met?
2. Are the expended costs within the budget estimate or contract costs?
3. Are the schedules of the program being met?
4. Is there any indication that the requirements, schedule, or cost goals will not be met?

The status reports to the lower levels of management involve a significant amount of detail, since most of the corrective action, in the event of any project difficulties, has to be taken by those lower levels. At progressively higher levels of management, progressively less detail is required.

Some companies conduct periodic review conferences during which the project manager personally presents the status of the project to assembled management officials. These oral presentations require the use of charts and other visual aids excerpted from the formal status report documents used in the organization.

The details of reporting forms vary among companies, but all the forms provide a way of reporting the same fundamental information. Whatever reporting system is being used on a particular project, the same general reporting means can be implemented. One fundamental type of report which is essential for providing the required status information is the management summary report. Other types of reports used to furnish different types of information discussions under PERT are the manpower loading display (Figure 11.5), the cost prediction report (Figure 11.6), the schedule prediction report (Figure 11.7), and the line of balance (Figure 11.8).

There are many other variations or types of reports that can be implemented, and it is up to the project manager to design and adapt any particular type deemed essential to the project. One thing that should be avoided is the temptation to adapt reports for a multitude of conditions, thereby building up a "paper mill" which will require so much effort in their completion that the program will suffer.

In addition to communicating down through the project organization or up into the management domain, the project manager must interface with organizations external to the company and with other company groups not associated with the project. Communications with the customer, with consultants, and with potential sub-contractors are some of the external dialogues that might take place. In addition, the project manager quite often must communicate with such other departments as accounting, legal, and personnel.

15.3   Instructions and Direction

The communication of instructions by its nature extends from the higher levels of authority downward. Most instructions, especially at the higher level of management, are issued verbally. There have been volumes of material written on the various aspects of instruction communication, and it is not the purpose of this text to discuss the philosophy or techniques of verbal communication, except to state that any form of communication must be sufficiently precise, complete, and lucid to guarantee that the maximum possible degree of understanding of what is being communicated exists among the parties involved.

Instructions relating to tasks, policies, and other formal matters must be documented. Even the verbal instructions mentioned above should be confirmed in writing if they are related to any matter that is significant. Whereas general instructions would take a normal narrative form, instructions of a technical nature would be issued according to a standard format. The use of a format guarantees that the basic necessary information is included.

The authority that is extended to the project manager to issue instructions and directions carries with it the responsibility for the results of such directions. An effective supervisor-subordinate relationship is essential for good management, but it should be noted that the project manager is given a responsibility requiring that a team effort be made by all individuals assigned to the project. Project managers must therefore exercise the fundamental principles of human relationships with their subordinates.

In giving direction to the project team, the project manager must make sure that the following major conditions are satisfied:

1. The direction will not lead to a deliverable item which does not comply with the contract specification.

2. The direction represents the most realistic and reasonable cost to the project budget.

3. The direction can be accomplished with existing or available personnel and resources.

4. The direction will not require that the planning and delivery schedule be compromised.

Any direction given by the project manager which affects any of the above conditions has an impact on the basic cost, delivery, or performance objectives of the project and probably necessitates some type of contractual amendment or revision to the planned in-house project. In the case of a matrix organization project, direction that involves the reassignment or rescheduling of resources must be checked with the functional manager, and if the project objectives are affected, the revised project plan is presented by both managers to the matrix executive for approval.

Whereas this text discusses the role of a project manager responsible for managing a contracted or in-house development effort, it should be noted that such using activities as government agencies and industrial organizations retain project heads who are vested with the responsibility to make sure that the project objectives of their respective employers are satisfied. The project head for the procuring or using activity and the project manager for the contractor have comparable responsibility and authority. It is essential that for a given contracted effort, the project leaders of both parties communicate frequently and cordially regarding the status of the contract. However, because of the inherent authority of the contractor project head, any direction, explicit or implicit, should be carefully considered and implemented so as to avoid contractual or legal complications.

For government and many types of commercial contracts, the principles discussed in Chapter 10 in relation to constructive change orders demonstrate the extent of diligence that project managers should exercise in giving direction to the outside world on their project.

15.4 Engineering and Programming Instructions

Identification of the different engineering and programming tasks is usually included in the WBS (work breakdown structure). The information derived from the WBS is used to create work packages which indicate how, when, where, and by whom a task is to be accomplished and what that task is. Thus the work package is a comprehensive document used for engineering and programming assignments to team leaders. In addition to the information noted above, the work package contains scheduling, budget, and similar information pertinent to the task. A variation of the work package is the task assignment form, which describes the objectives of the engineering or programming effort.

Traditionally, the instruction which constitutes a detailed breakdown of the work package or task assignment for the design of hardware is the EO (engineering order).

The issuance of the EO does not constitute an end in itself. Although the EO attempts to convey all the required information in the clearest and most complete manner possible, questions of interpretation and explanation always evolve. Thus the project design engineer and the various group leaders to whom the EO is directed will invariably engage in many discussions on the assignment. If a problem arises that cannot be resolved by the project design engineer, it is referred to the project manager, and the working-level communication cycle illustrated in Figure 15.1 is traversed.

During the course of a major project, changes are almost inevitable. A change may be initiated internally or may be initiated by the customer. At any rate, any change must be reflected in revision of the design, and appropriate information must be transmitted to the design engineers. The means used to incorporate a change in design is the EC (engineering change form). The EC provides all the technical, cost, and schedule information desired for revising the product in question.

As in the case of the engineering of hardware, instructions for the different tasks involved in software design are derived from the decomposition of the WBS and are described in work packages. A further breakdown into computer software or program modules is provided in module instructions (or module folders, etc.). Each module instruction must perform a specific function, be complete within itself as far as testing and changes are concerned, and constitute an element that is readily integrated into the whole software system. It is the responsibility of the project manager to have the computer program system fractured into modules that can be readily monitored by the management system being used (such as PERT), that can be identified in the configuration management system for the project, and that are self-contained in that they do not require revision of other modules.

15.5 Quality Control Instructions

Each project has specific performance objectives which are generally quantitatively expressed in terms of accuracies, reliability, response speeds, etc. In order to achieve the objectives, quality control criteria (or standards and procedures for certifying that the objectives are being met) are issued as quality control instruction.

In the case of the RLM simulator, one of the performance objectives is ability of the system to present a display of the areas illuminated by the simulated radar as the gaming area is traversed. Thus the system must be able to retrieve the stored terrain data; perform calculations relating to ranges, attenuation, shadow effects, etc., in real time; and present the information as a display that duplicates what would be observed in an actual mission over the area that is simulated. The above can be translated into the following functions of the subsystems shown in Figure 3.3:

1. The radar characteristics controller must be capable of instant response to the signals depicting the simulated aircraft flight and antenna scan.

2. The controller must have instant access to the applicable terrain data stored in the data storage unit.

3. The data processor unit must be capable of performing the digital calcu-

lations relating to ranges, etc., that are necessary to achieving the required real time display.

The hardware and software have been designed to achieve the performance objectives described. It is now the responsibility of the quality assurance group to establish the criteria and to make sure, by inspection testing, etc., that the quality of the hardware and software is at a level of excellence that will result in the achievement of the design objectives.

The project manager has final responsibility for implementing the quality control program. The execution of the program is usually assigned to a quality control coordinator, and the instructions that are distributed are identified as quality control orders (QCOs).

15.6   Drafting Instructions

There are two basic schools of thought about the organizational status of the drafting effort. One holds that since drafting is so closely related to engineering, the effort should be integrated with the engineering department itself. The other holds that all drafting should be accomplished in a separate organization. There are many arguments that can be advanced for each concept. Most companies, especially the larger organizations, have adopted the separate-department concept. The argument for the separate drafting department is that by having a nucleus of experienced draftsmen, standardization of drafting procedures and its efficiencies can be promoted, and such an organization minimizes the amount of time that creative engineering talent must otherwise spend in routine drafting efforts.

In Figure 1.2 it can be seen that the drafting department is organized as a separate organizational unit under the head of engineering. The lines of communication are routed from the design engineering groups to the chief draftsman via the project design engineer. The same lines of communication for drafting instructions are applicable to the matrix organization, except that in the matrix the discipline supervisors become involved.

The communication between the design engineering groups and the drafting personnel is much less formal than in other parts of the organization because of the unique form of information that is transmitted. The draftsman is in effect translating into formal engineering drawings the ideas and concepts that the design engineer creates. The documentation by the design engineer usually exists in the form of sketches and notes which must be verbally explained to the individual draftsman doing the work.

However, in spite of the nature of the task, formal instructions must be given at the start of any drafting task for planning and scheduling purposes. Therefore the engineering department issues some form of drafting order (DO) to the chief draftsman. The DO contains pertinent information that the design engineer must document so that the drafting department may fully understand what must be contained in the drawings. The DO includes the following information: description of the task, supplementary sketches and other documents, budget hours allowed for completion, and the date that drawings are required.

After receiving the DO, the chief draftsman can schedule the work and provide the detailed information required for the assigned draftsman to accomplish the task.

Any revision or change in the design must be transmitted as an instruction to the chief draftsman and would be contained in a drafting order change document.

15.7 Purchasing Instructions

A requirement for purchasing is originated by the project design group, submitted to the project manager for approval, and formally issued as a requisition for purchase to be executed by the project purchasing agent.

The factors that must be considered in reaching a decision on whether to make or buy a component or subsystem were discussed in Section 12.5. When the decision to buy is made, the requisition, which contains complete detailed information with specifications, quantities, required date, suggested vendors, etc., is routed to the reliability control department and other interested departments to make sure that the item will meet the standards necessary for its application. The requisition is then routed to the project purchasing agent.

Once the purchase order is placed, copies of the document which give all the pertinent information relating to the purchase are sent to the project manager and to the lead design engineer or the discipline engineer. A check must be made by the design engineering department for verification that the part being ordered does, in fact, meet the specified engineering, reliability, and other requirements. Subsequent to the placing of the purchase order, changes in delivery, design, etc., may become necessary which require revision of the purchase order; changes in the communication channels, corresponding to those in the purchase order, would also be necessary.

15.8 Production Instructions

The creation of a hardware design and its translation into a tangible piece of equipment involve two distinct types of effort: engineering and production. In the creation of software the line separating design (engineering) and coding (production) is less distinct, since the efforts are often performed by the same team of programming personnel. Furthermore, when configuration management disciplines are not rigorously implemented, changes in the design to accommodate the coding effort often become necessary, complicating the orderly creation of the software. Current management techniques involve use of comparable formal procedures for producing hardware and software. Program writing, which includes coding, is generally treated as the production phase of the computer program.

In the creation of hardware, the monitoring of the production and assembly of a complex piece of equipment involves many disciplines and controls. The project manager becomes involved in production details only when an unusual problem arises which might jeopardize the entire project. Basically, project managers are concerned only with the production schedule, quality, and cost, and these are the factors covered in reports to them.

The initial instructions issued to the project production coordinator are general in nature and in essence direct that the necessary planning be made in accordance with the analysis and information derived and used for the technical proposal that resulted in the contract award. The document used to convey this instruction is sometimes referred to as the PPO (project production order). In the case of the RLM simulator the PPO would be supplemented by the specifications, copies of the technical proposal, and other related documents to provide all the necessary information.

After receiving the PPO, the project production coordinator breaks down the project into the various necessary tasks, such as sheet metal work, cable assemblies, and machining operations, and lays out an estimated schedule based on the date when inputs from engineering are expected.

When the engineering design is established and the drafting department has completed the engineering and manufacturing drawings of a particular component or assembly, the information is incorporated into a PWO (production work order) and transmitted to the production coordinator for further transmission to the particular department that will fabricate and assemble the part.

The PWO is routed through the project manager, who authorizes the work and verifies that the indicated schedule of work does conform to the master schedule of the project and that tolerances and quality requirements are satisfactory. Periodically, the project production coordinator transmits a status report relating this information to the project manager.

The coding requirements included in documents generally used for communicating instructions in the production of the computer program are contained in the PDD (program description document). The PDD includes the specific design details of the program as derived from the PPS (proposal performance specification) and the PDS (program design specification) and provides the information necessary for the coding effort for each program module. The modules described in the PDD were derived from the work breakdown structure. The PDD includes flowcharts to illustrate processing logic and information relating to interfaces. The PDD for the software is usually considered to be the counterpart of the PWD for hardware in communicating production instructions.

15.9   Planning and Control Communications

In line organizations, the planning and control functions, which include monitoring the scheduling and cost control of the project, are performed by a group reporting as line or staff to the project manager.

The responsible head of the planning and control effort is called the coordinator. The coordinator's responsibilities include gathering information regarding the project cost and schedule status, comparing the information with the planned and budgeted data, and submitting status reports to the project manager.

The planning and control coordinator receives periodic status reports from the head design engineer, the production manager, the quality coordinator, and other heads. The information so received is then collected and presented in a formal narrative report and is graphically depicted in such displays as the manpower loading display (Figure 11.5), the cost prediction report (Figure 11.6), the schedule prediction report (Figure 11.7), the LOB report (Figure 11.8), and any other similar documents that are deemed necessary by the project manager.

In addition to providing the statistical information discussed above, the coordinator must provide information relative to areas where difficulties exist and describe what factors are responsible for the troubles. In addition, the coordinator provides remedial actions wherever possible as suggested by the department heads. In this manner, the project manager obtains a complete briefing on the project and is in a position to make a comprehensive, intelligent, and constructive report to higher levels of management.

In the matrix organization, the discipline supervisor who performs the planning and control functions must communicate project status information and receive instructions from two bosses: the project manager and the functional manager. As in the line organization, the discipline supervisor responsible for tracking the status of the project is delegated the authority to gather pertinent data from the other discipline supervisors of the matrix for the reports to be conveyed to the matrix managers.

15.10   Communication with the Customer

The project manager in both the line and the matrix is also the focal point for communications outside of the organization. This is the individual to whom the customer directs any inquiries, instructions, or suggestions related to the project. Very often, a company employs a marketing director whose primary functions are maintaining good relations with customers and acting as an intelligence source for obtaining various types of procurement information. Although the marketing director may be the official contact for a customer, practically all questions are passed on to the project manager, who provides the answers and usually ends up as the contact.

The major forms of communication between the project manager or the marketing director and the customer are (1) letters, dispatches, and other forms of the formal written word, (2) telephone conversations, (3) personal meetings relating to the monitoring effort, and (4) program review meetings on the project.

It is mandatory that the project manager see to it that any written communication requiring a reply is promptly answered. Prompt replies to letters and other written communications are vital for the following major reasons:

1. A prompt reply minimizes any delays in the program progress by contributing to prompt decision and action.

2. A failure to answer a communication within a specific period (usually 30 days) is assumed to constitute concurrence with the point in question, and the addressee, by remaining silent, is assumed to have given tacit approval to the customer's point of view. Contractually speaking, then, the rights of the project manager will be lost on any issue in a letter which was not answered in time.

Another vital point relates to verbal discussion, such as telephone conversations and meetings. Written records of the content of all such communications should be kept. In the case of meetings, minutes usually are published and distributed. For other types of communication, a written record of any conversation is mandatory. If the conversation is related to any decisions, action items, or other significant matters, the project manager should immediately draft a letter to the other party confirming what was discussed and detailing the agreements and action items.

In the matrix, the project manager is required to coordinate with the functional manager any actions that are required as a result of communication with the customer. If there is a conflict or if a change in project objectives is required, the matrix executive becomes involved. However, in general, the project manager represents the point of contact for the project.

15.11 Decision Making

The fundamental purpose of providing for channels, methods, and tools of communication is to permit responsible management officials to arrive at proper decisions. At this point, the fundamentals involved in arriving at a decision and how the decision-making function is contingent upon communications will be discussed.

1. Identifying the existence of a problem is usually facilitated by the fact that some objective is not being achieved. For instance, the effort or cost to complete the design for the radar characteristics controller may be exceeding the budget. Investigation might reveal that the cause of the problem is a particular design flaw. It is essential that the problem be defined as precisely as possible.

2. The accumulation and evaluation of the facts related to the problem constitute the second phase in deriving a decision. In the case of the design problem cited above, the requirements of the specifications and the specific technical requirements, as well as the difficulties which hinder the achievement of these requirements, must be identified and evaluated.

3. Consultation with other individual experts in the area in which the difficulties exist is the third phase of the decision-making pattern. The introduction of outsiders to consider the problem brings into the picture new points of view which may result in a solution that was overlooked by the individuals "living with" the problem. The least that the outsiders may offer is an observation that the existing design approach provides no solution and that an alternative approach is needed.

4. The fourth phase, suggested above, is to seek alternative approaches if the existing approach is judged to offer no solution. If, in fact, the existing path of action is deemed to be unfeasible, the earlier a new approach is adopted, the less the schedule slippage and wasted effort.

5. The fifth phase is to analyze thoroughly the proposed solution to the existing problem or to evaluate in detail any alternative approach to make certain that whatever course of action is taken, the result will be successful.

6. The last phase is the actual decision itself and the action taken to implement the solution to the problem.

Whereas the mechanics of identifying the problem, gathering information, etc., are essential to fulfillment of the decision-making responsibilities of the project manager, the actual execution of the decision requires mental disciplines that should also be recognized. A decision must not be influenced by any emotions or personality conflicts. Therefore the data relating to the problem must be screened to establish its relevancy. Figure 15.3 illustrates the types of data that are usually accumulated for a problem and the amount of data that is normally considered relevant. Project managers must, with complete objectivity, identify and discard the irrelevant and erroneous data so that they can concentrate only on the relevant information in making their decision. It should be

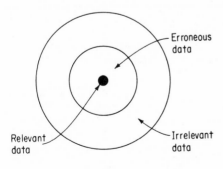

Erroneous
data

Relevant
data

Irrelevant
data

FIG. 15.3 Types of data gathering for consideration of a decision.

noted that if any erroneous or irrelevant information is used as a basis for a decision, the logical conclusion that can be reached is that the decision itself will be in error.

15.12   Summary

The successful monitoring and rendering of decisions on a project are dependent on the effectiveness of the communication system. The project manager in the line organization is the key person in a project and is dependent on the prompt receipt of accurate, complete, and precise information. In the matrix, the project manager is also the key contact for project communications, but must share information with the functional manager to update the project status and to work out courses of action.

The two types of communication are routine periodic status reports and special communications initiated by the presence of a problem. In addition to receiving communication from the working levels, the project manager must transmit information to higher management. The receipt of communication forms the basis for decisions which are translated into instructions on the project.

The type and form of the instructions provided to the department heads in the line organization and the discipline supervisors in the matrix varies for the type of project tasks that are being performed. However, the common denominator for all instructions and communications is the achievement of the cost, schedule, and performance objectives of the project.

The management echelons of a company are primarily interested in whether the specification requirements, estimated costs, and contract delivery schedules are being met, since all three objectives must be met for a project to be successful.

The tools of communication vary in detail in different companies, but essentially are all designed to convey the general program status, the cost status, and the delivery schedule.

BIBLIOGRAPHY

AMC Regulation 11–16: Planning and Control Guide, Headquarters, U.S. Army Materiel Command, Washington, D.C., 1963.

Newell, Wendell, and Albert Goldstein: *Software Management of a Complex Weapon System Simulator from Technical Report, NAVTRAEQUIPCEN IH 316,* U.S. Naval Training Equipment Center, Orlando, Fla., 1979.

Rantl, Robert M.: *Guidelines to Productive Management,* Management Review, 1979, New York, N.Y.

16

RELIABILITY AND
MAINTAINABILITY

16.1 Background

The technological advances incorporated in the equipment used for military, industrial, and consumer applications have created serious problems in maintaining reliable and stable operation over even nominal periods of time. Because of the demands by different users for highly automated equipment, designers and manufacturers are forced to emphasize reliability in the equipment that they produce.

There are a multitude of factors that enter into consideration of reliability. The determination of reliability is in the final analysis dependent on whether a particular component, element, or module is capable of performing its specific function without degradation or failure for a specific period of time under the conditions for which it was intended.

The primary environmental factors which affect the reliability of a system and which must be considered in the design of the equipment include the following:

1. *Temperature:* Range in which the system must function in an acceptable manner
2. *Vibration:* Magnitude and frequency of vibration forces under which the equipment will perform

3. *Shock:* Force and repetition rate of shock that the equipment can withstand

4. *Operating period:* Length of time over which the equipment can function without exceeding a specified number and type of failures

With the introduction of new technology, the emphasis on the different environmental factors shifted and design features for assuming reliability were revised. Whereas vibration and shock were major reliability considerations for the old vacuum tube designs, temperature became a major parameter that affected the functions of solid-state components such as transistors. System designs based on the application of the new technology of integrated circuits and other "chip" forms permitted high-density equipment with an inherently high reliability. In the design of integrated circuit systems, the achievement of acceptable reliability depends primarily upon the judicious selection and arrangement of circuit elements that make up the subsystems.

For digital computer systems, not only must the processor and its associated peripheral systems be capable of meeting reliability requirements, the computer software itself must possess unique reliability qualities. Whereas in hardware systems, major causes of reliability failures are wear, heat, environmental exposure, improperly designed circuits, etc., the reliability problems in computer software are primarily the result of programming and coding errors. Thus a different emphasis is needed in the creation of hardware than in the creation of software to achieve acceptable reliability characteristics in each case.

Closely akin to the reliability parameter are the factors of maintainability, availability, and effectiveness. In addition to reliability, maintainability (or the ability to expeditiously make necessary repairs or corrections when some failure or error occurs) is also of vital concern to users of a system. In the same family of parameters there is also the factor of availability, which is a function of reliability and maintainability. Effectiveness, which is a measure of the combined factors of reliability and availability, is another factor relating to the utilization of a system.

16.2 Reliability Concepts

Reliability is a statistical term generally defined as the probability that for a specific period of time and under specified conditions a piece of equipment or a system will perform in a manner that is acceptable. It should be noted that when reliability is incorporated as a contractual requirement, the terms of time, conditions, and acceptable performance must be expressed quantitatively. A detailed study of reliability involves statistics and mathematical derivations which go beyond the depth of this test. However, the primary notions of reliability, such as mean time between failures (MTBF) and availability, are essential to the work of the project manager.

The ultimate in reliability is the ability to have the equipment operate as specified without failures over a specified operating period during the life of the equipment. Modern radios, television sets, and steam irons are examples of equipment in which an unusually high degree of reliability has been realized. The average television set gives consistently satisfactory performance for years without any breakdown. The degree of reliability achieved with consumer products is the result of exhaustive tests and redesign before the product is released for manufacture.

Whenever circumstances permit, the procurement of military systems involves testing of prototype equipment and implementing whatever design revisions are required to achieve the specified performance such as reliability required of manufacturers. For computer systems, software programs that are used to diagnose problems and isolate circuits failures enhance the maintainability capabilities of the system.

It is interesting to note that because of escalating costs for repair services, many companies are designing their products in such a way as to achieve a higher degree of maintainability. Examples are the use of throw-away circuit cards for television sets and similar features for various electronic and mechanical devices which are being promoted by various manufacturers.

The simulator specification, for example, requires that the trainer be capable of meeting an MTBF reliability of 100 hours and an MTR (mean time to repair) maintainability of 1 hour for a continuous 24-hour operating period. The design engineer must convert these requirements into subsystem design criteria and component selection.

The question might be raised as to what the difference in design approach would be if the requirement were for a 48-hour continuous operation instead of a 24-hour period.

The answer to such a question is apparent after consideration is given to the question of reliability. The degree of reliability is ultimately based on the following two major requirements: (1) the ability of the individual components and elements to perform within their individual specified tolerances and conditions for their rated life and (2) the ability of the design of the different subsystems to function as intended under different conditions.

The selection of the individual components is based on the MTBF values. What constitutes an acceptable mean time between failures is dictated by the specified operation time (e.g., 24 or 48 hours) and a distribution of the total of all the MTBF values of components in a system or array of equipment. The statistical distribution of the data will result in a probability curve from which a prediction of reliability can be derived.

The other aspect of reliability is how well the design of the subsystem can maintain the specification requirements for performance under the range of conditions to which the equipment would be subjected. This requirement is achieved by networks, feedback circuits, and various other design features that will automatically compensate for adverse effects due to changes in environmental conditions.

The degree of reliability of performance of any newly developed piece of equipment depends on the period in the life of the equipment when the test for reliability is being made. Figure 16.1 shows a typical reliability curve that one might expect during the three time periods of the life of a system.

The degree of reliability realized during the equipment infancy is poor due to the fact that the so-called shakedown period exposes major and minor deficiencies that must be corrected. A large percentage of these deficiencies cause a breakdown or degradation in the operation of the equipment.

After shakedown, the normal degree of reliability of the equipment is experienced during the productive period which makes up the major portion of the life of the equipment. The reliability requirements and design objectives are expected to be realized during this period.

At the end of the life of the equipment, obsolescence begins to affect the reliability of performance at an increasing rate. During this period, failures occur due to the literal wearing out of parts and components. In Figure 16.1, the reliability curve for the equipment shows a decreasing period of time between breakdowns of the equipment until a breakdown occurs every 4 hours, which renders the utilization capability of the equipment unacceptable.

At this point, the equipment must be either completely overhauled and refurbished or scrapped. The situation is identical to that which confronts the automobile owner when, after the vehicle has been driven a certain number of miles, breakdowns and subsequent repairs are necessary, and ultimately the frequency of breakdowns renders the automobile useless for the owner. When reliability is discussed, it is to be assumed that the reliability results are derived

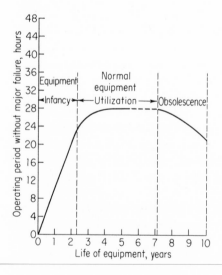

FIG. 16.1 Equipment reliability curve during utilization lifetime.

from operation of the equipment during its normal equipment utilization period or during the center portion of the curve illustrated in Figure 16.1.

16.3 Reliability Qualities in Systems

To satisfy the requirements for highly sophisticated and automated military and industrial systems, complex equipment must be developed. It would have been almost impossible to design modern systems with acceptable reliability qualities with the technology of even the recent past. Fortunately, the use of solid-state elements in integrated circuits, VLSI (very large-scale integration) technology, and other state-of-the-art advances have imparted high degrees of reliability which more than compensate for the reliability problems that would otherwise have been presented by the increasing complexity and size of modern systems.

Although the trend for contemporary systems is toward designs utilizing integrated circuits or chips, use of transistors, resistors, capacitors, etc., continues to be widespread. Project managers must adopt procedures to make sure that the system is designed to ensure acceptable reliability features for any type of technology used. Therefore, a reliability analysis is required by design and reliability engineers of the functional, reliability, and maintainability requirements of the system to establish design approaches, to select appropriate parts and components, and to identify diagnostic features for troubleshooting.

16.4 Temperature and Stress Factors

In traditional electronic circuit design, temperature variation is handled by properly selected components and self-compensation circuits which usually take the form of feedback networks. Most electrical or electronic components change values as their temperature changes. A resistor experiences a certain value when initially energized, but during operation of the equipment, its temperature changes as well as its value. If the function of the resistor in the circuit is critical, some compensating feature must be included in the design in order to achieve the desired degree of stability.

Temperature is very important in circuits where transistors are used. The effective junction temperature of a transistor depends on the ambient temperature and the junction temperature rise due to the flow of current in the transistor junction. The characteristics of a transistor are such that its functions in a circuit will vary with temperature, and if such variations in the output of a particular transistor, for instance, would cause or contribute to a degradation of equipment performance to the degree that the equipment would be rendered unreliable, then a compensating design measure such as a feedback circuit to maintain suitable operation over a temperature range is necessary. The design engineer must recognize which components and circuits are critical, analyze the output of a critical component over temperature ranges, and design the circuit to compensate for the undesirable variations. The project manager should be knowledgeable of the critical areas of a design and should be prepared to direct design action where necessary in order to achieve the required reliability objectives.

The specification and selection of components that will permit the realization of the reliability requirements of a subsystem make up a second and probably more important factor of reliability design. Before the selection of a component can be made, the engineer must establish the criteria for the selection. Thus the reliability analysis of the particular component under consideration must be interpreted and the individual component specification established. With the specification established, the component can be fabricated or purchased.

In examining any particular component of a design to establish its reliability specification, the design engineer must determine what type of component malfunction would render the equipment inoperative or result in a catastrophic breakdown.

The normal concept of failure is the obvious breakdown, such as that resulting from the snapping of the drive shaft of an engine. For electrical equipment, failures usually occur in two models—either an open or a short. The design engineer must then determine for the component under analysis whether the occurrence of a short or an open or either type of component failure would result in a catastrophic failure of the equipment. If it is estimated from the specification of the component and the application in the circuit that the probability of component failure due to a short is 70 percent and that the probability due to an open is 30 percent, and if the failure of the component will result in catastrophic equipment failure only when the component failure is a short, then the 70-percent figure would be used in the reliability design.

One facet of component reliability is the failure rate expressed as the rate per specific number of hours. Another method of expressing failure rate is the expected life expressed in hours under rated design conditions. The period of life of different components in a subsystem would be scaled to a common base for purposes of calculation. The expected life of a particular resistor in a circuit might be 20,000 hours. This can be expressed as a failure rate of 1 in 20,000 hours. A transistor having an expected life of 10,000 hours would then have a failure rate expressed as 2 in 20,000 hours.

In a calculation of the effective failure rate, the failure rate of a component which is based on design conditions of operation such as rated load and specific ambient temperatures must be modified by factors which reflect the operating conditions of the subsystem design that would utilize the component. In reliability engineering, these factors are referred to as stress conditions, which fall into two categories, namely, the maximum-limit and time-function conditions.

Failures as the result of maximum-limit conditions occur when the stress imposed on a component reaches the maximum limit of the component capability. The failures resulting from the time-function conditions occur more or less as the result of fatigue after a period of time.

For the electrical components being analyzed on a subsystem, the maximum-limit conditions would generally be stresses such as voltage, current, vibration, and shock. The time-function conditions would involve temperature, current as it relates to heat-producing characteristics, fatigue due to vibration, and, in a general way, the aging process. The failures due to time-function conditions will occur in decreasing periods of time as the level of the stress is increased.

The design and selection of a component which will meet the maximum-limit stress conditions are straightforward. This effort requires that the engineer calculate all possible sources of stresses and determine how to withstand the worst possible condition. Basically, this involves the traditional engineering analysis and the application of a suitable safety factor. The same principle would apply whether a bridge or a radio receiver were being designed.

The time-function stresses involve a special analysis for reliability which must include consideration of the magnitude of a stress, the failure mode that would result in a catastrophic failure, the mean time between breakdown, and other factors.

Table 16.1 illustrates how the analysis of various factors would be tabulated for the application of various components of the subtractor system design. A failure of resistor R1, for instance, would always be an open, so that the probability of a short occurring would be zero and an open would occur 100 percent of the time. The effect of an open for R1 would be a failure. Capacitor C3, if it failed, would experience a short 40 percent and an open 60 percent of the time. In other words, if a failure occurred for C3, the probability of a short occurring would be 4 out of 10 and the probability of an opening occurring would be 6 out of 10. A failure of C3 as a short would render the subtractor system useless or be catastrophic (indicated as an F), whereas a failure of C3 as an open would be trivial and not affect the system (indicated as an N).

TABLE 16.1 Tabulation of Effects of Component Failures on an Electronic Circuit

Component	Short		Open	
	Probability, %	Effect on system	Probability, %	Effect on system
Resistor R1	0	—	100	F
Resistor R5	0	—	100	F
Capacitor C3	40	F	60	N
Capacitor C7	80	F	20	F
Capacitor C8	70	N	30	F
Mode D6	50	F	50	F
Mode D7	50	N	50	F
Transistor Q6	50	N	50	F
Transistor Q7	60	F	40	F
Relay contact R2	10	F	60	N
Relay contact R2	10	F	20	F

F = catastrophic failure; N = failure that will not affect the system.

The information in Table 16.1 establishes the estimated statistics regarding the relative impact of failures of different components on the trainer reliability. The design and reliability engineers must now make an estimate of the probability of failure of each component, or, expressed a different way, of the expected failure rate over a particular period of time.

A vast amount of data has to be collected relating to the results experienced with running life tests on various types of components, such as resistors, capacitors, and transistors. For instance, results of tests on transistor life while operating at different temperatures are summarized in Table 16.2.

The stress factor or ratio relates to the percentage of rated load to which a component is subjected. For instance, a transistor delivering twice its rated load may have a life expectancy of one-tenth normal life. In like manner, a transistor operating at one-half its rated load may experience a life expectancy of five times its normal life. The stress ratio, expressed as the operating load over the rated load, is important in calculating the life expectancy or failure rate of a component. For application to the calculation of net failure rate, the stress factor is converted to a figure identified as the stress life factor. For the transistor in the above example, whereas the stress ratio is 0.50, indicating operation at one-half rated load, the life would be increased by a factor of 5, or the failure rate would be reduced to one-fifth the normal rate. Thus the stress life factor would be 0.20.

Table 16.3 illustrates a tabulation of the main reliability factors and the simplified application of the various reliability factors to derive the net failure rate of each type of component. As indicated in the table, the net failure rate is derived as the product of the factors in columns 1 through 5. By calculating the sum of failure rates and dividing the time base by this sum, the meantime between failures of the subtractor circuit would be established.

The net failure rate of the complete system can be estimated by making a reliability prediction of each individual subsystem and deriving the failure rate or expected life between failures of the equipment itself as a statistical composite of the failure rates of each subsystem. If, for example, there were 10 subsystems of a piece of equipment, each of which had a meantime between failures of 2,000 hours, then by the most elementary and crudest calculation, the meantime between failures would be 2,000/10 hours or 200 hours. If the

TABLE 16.2 Summary of Typical Transistor Life at Different Operating Temperatures

Temperature, °C	Life, hours
200	2
150	200
100	4,000

TABLE 16.3   Tabulation of Factors for Reliability Prediction for an Electronic Circuit

Part	(1) Quantity	Stress ratio	(2) Stress life	(3) Temperature factor	(4) Failure rate per 10,000 hours	(5) Probability factor	(6) Net failure rate per 10,000 hours
Resistor R1	5	0.5	0.2	0.7	0.8	1.0	0.56
Resistor R2	1	0.7	0.3	0.6	0.7	1.0	0.13
Capacitor C3	2	1.0	1.0	0.8	0.9	0.4	0.58
Capacitor C7	1	0.6	0.3	0.7	1.2	1.0	0.14
Capacitor C8	4	0.5	0.2	0.7	1.2	0.3	0.19
Diode D6	1	0.3	0.1	0.6	1.9	1.0	0.11
Diode D7	3	1.0	1.0	0.6	2.2	0.5	1.98
Transistor Q6	2	0.7	0.4	0.5	2.7	0.5	0.64
Transistor Q7	4	0.5	0.3	0.7	2.7	1.0	0.23
Relay coil	1	0.8	0.7	0.8	0.6	0.7	0.19
Relay contact	1	0.8	0.7	0.8	1.0	0.3	0.22
							4.97

Total failure rate: 4.97/10,000 hours
Mean time between failures: 2,000 hours

equipment was comprised of 20 subsystems, each with a meantime between failures of 2,000 hours, the reliability figure of the equipment would be 100 hours. The general conclusion is that as the complexity of an equipment goes up, the reliability factor goes down. All other factors are kept the same. Another conclusion that can be reached is that if a high reliability is desired in a complex piece of equipment, some of the reliability factors listed in Table 16.3 can be improved. For instance, the stress ratio can be lowered by operating a component at a lower percentage of its rated capacity, which would result in a longer expected life for the component.

16.5   Integrated Circuits and Software

Even though integrated circuits possess significantly higher reliability than components based on earlier technologies, the larger numbers of circuits and the higher reliability objectives require that the greatest number of identical circuits be used and that the optimum grouping of parts be achieved. By minimizing the number of different circuits, the number of the different failure mechanisms possessed by the array of circuits is minimized. The effective grouping enhances troubleshooting procedures when failures occur.

The number and variety of functions that integrated circuits available in the market can perform permit their use in designing almost any type of subsystem. Thus the engineer designing a system becomes heavily involved with design logistics, or the art of selecting the array of integrated circuits which will perform the necessary functions and provide optimum reliability capabilities.

In selecting the different integrated circuits for a subsystem, the design engineer must make sure that the basic features of proper element design, good workmanship, and adequate derating are provided. The goal is to achieve the highest possible degree of homogeneity in a system; this in turn makes it possible to anticipate the types of failure that will afflict families of circuits, and to provide for redundant elements, for troubleshooting aids, and, in general, for a feature of simplicity in a complex system.

In establishing the policies and guidelines for the effort, the question that confronts the project manager concerns what criteria are available to identify the integrated circuits that are selected. Manufacturers of such products provide technical and performance descriptions of their products which include information relating to reliability capabilities. In addition, industrial and government organizations constantly perform objective evaluations of different products, and the results are published to provide potential users with the data for selecting items which best fit the desired applications. The most comprehensive evaluation of such parts as integrated circuits is performed by the U.S. government. Specification standards have been established, and integrated circuit parts that have been tested and have successfully met the standards are included in government documents such as the QPL (qualified parts list) and other catalogues of government-approved parts. Thus by taking advantage of available technical data and exercising logistics design functions, the designer can provide systems with optimum reliability.

The reliability of the digital computer can be enhanced through the design of the computer program. Error checking of memory data or incoming data can reveal reliability difficulties. In addition, features such as error correction can be designed and programmed into the system to improve reliability. Executive programs can be used for routine maintenance checks of the system to uncover potential problems prior to their occurrence in the operational mode.

A computer program is considered to possess acceptable reliability if it can perform its specified functions without significant errors. In effect, a high reliability for software is accomplished primarily in a comprehensive test effort during the development of the program. The organized testing of hardware and software represents the most effective means of making sure that the desired reliability of a system under development will be realized.

The achievement of reliability for any piece of equipment is a task that continues even while the equipment is in operation in the field. Reports of failures or troubles are usually referred to the reliability group for analysis and correlation. If such reports indicate a recurrence of a particular type of difficulty, a design analysis is made and a modification to the equipment is implemented to correct the difficulty and thereby achieve the desired improvement in reliability.

16.6 Maintainability Principles

The maintainability of a piece of equipment is a measure of its ability to be repaired expeditiously when failure occurs. The quantitative measure of maintainability is usually referred to as MTR. The term is usually further defined to establish the nature of the failure in the equipment—e.g., is there a spare part for the component that failed?

The design for maintainability requires that basic features such as access to the inner components be provided, test points be identified, diagnostic programs be provided, instrumentation and tools be available, an assembly tester and power and air pressure supplies be provided, adequate spares be stocked, etc. In other words, the equipment and software design and the facilities must be such that they provide optimum capability to make repairs or corrections and to get the system to function normally after a failure occurs.

A significant parameter which establishes the measure to which the equipment is capable of being utilized is the availability factor. Availability is expressed as a decimal and is established by the values of MTBF and MTR. The availability factor is determined by the following relation:

$$\text{Availability A} = \frac{\text{MTBF}}{\text{MTBF} + \text{MTR}}$$

Expressed in another way,

$$\text{Availability} = \frac{\text{Operating period}}{\text{Operating period} + \text{repair downtime}}$$

For the RLM simulator the specified requirements for MTBF and MTR are noted to be as follows:

MTBF (mean time between failures) = 100 hours

MTR (mean time to repair) = 1 hour

Therefore the availability factor of the trainer would result in a value of

$$A = \frac{100}{100 + 1} = 0.99$$

The effectiveness factor is a measure of the probability that the equipment will be ready and capable of performing its function and that it will not experience failure during its mission period. It is determined by the value of the equipment reliability, the value of the availability, and indirectly the ability to effect repairs. For a predetermined level of performance, effectiveness can be expressed as the product of reliability and availability.

Effectiveness = R × A

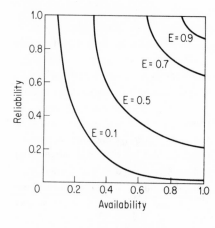

FIG. 16.2 **Effectiveness curve for values of reliability and availability.**

Because of the qualifications related to a specific mission, as noted above, the above expression for effectiveness should be used only when the conditions are preestablished. Effectiveness is generally used for military systems and equipment where equipment performance at a particular time is of vital concern.

Figure 16.2 illustrates a typical family of effectiveness curves for a piece of equipment that might be designed for different values of reliability and availability.

16.7 Summary

The general definition of reliability is the ability of the equipment to operate under specified conditions for a specified period of time without interruption due to equipment failure or degradation of performance. The specific defintions of the length of time that the equipment is to operate and what constitutes equipment failure are dependent upon the type, function, and design of the equipment to which the reliability is applicable.

The reliability of any equipment is based on the ability of individual components to perform as required and on the ability of the design of the various critical subsystems to compensate for changes in load, temperature, and other conditions that vary during equipment operation. The degree of reliability will differ, depending on what period in the equipment life reliability is observed. The three basic periods are infancy, normal utilization span, and obsolescence. Due to different factors, relatively poor reliability is experienced during the infancy and obsolescence periods.

The basic measuring tool of reliability is the MTBF. The two modes of reliability failure are catastrophic failures causing the shutdown of a system and degradative failures in which the equipment suffers due to progressive malfunctions causing a loss of accuracy or stability.

The major environmental factors that affect the reliability of systems are temperature, shock, and vibration. Systems designed with transistors and similar solid-state components require that special features to minimize the effects of high temperatures be provided. Since the more recently developed integrated-circuit modules inherently possess high reliability capabilities, the task of design engineers working on such modules is to make sure as few different types of high-quality integrated elements are incorporated in the system design as possible. Thus the emphasis of the engineers' task shifts from design disciplines to judicious parts selection when a system using integrated circuits is under development.

Reliability of computer software is a function of how well the software is designed and tested. The reliability of a computer system is enhanced by designing programs that include error-checking and error-correcting capabilities.

The implementation of reliability is accomplished by a special reliability and quality-control group which, in the case of the RLM system, is headed by the project quality coordinator. A reliability engineer is generally assigned to advise the engineer designing a particular subsystem on reliability requirements. The reliability requirements for the system are identified in the reliability analysis. Each system is analyzed to determine which components are critical to the operation of the system, what type of failure of each component would affect system operation, and what the probability is of each particular failure occurring.

A further analysis of each component is made in a reliability prediction study. The reliability prediction study considers such factors as the quantity of each component used, the stress ratio, the temperature factor, the failure rate per period of time, and the probability factor mentioned above. As a result of the reliability prediction analysis, the estimated MTBFs of the system can be derived from a rough statistical operation.

Maintainability is a measure of the ability to repair a piece of equipment when failure occurs.

Availability is a measure of the utilization capability of a piece of equipment and is established by the equipment reliability and maintainability.

For a specified set of conditions, the factor of effectiveness, which is the product of reliability and availability, can be established.

BIBLIOGRAPHY

Depp, R. E., and John Raye: *Reliability Enhancement of Simulators Through Parts Control: Interservice/Industry Training Equipment Conference Proceedings*, Report 1H-316, Naval Training Equipment Center, Orlando, Fla., 1979.

PRODUCTION AND QUALITY CONTROL OF HARDWARE AND SOFTWARE

17.1 Production of Development Systems

Production as a subject covers a very broad area and involves a multitude of concepts and disciplines. The assembly line production of a toaster is vastly different in its control and organization from the production of a custom-built reactor for a power plant. Since the production of a prototype device necessitates a close liaison with engineering, project managers require lines of communication with both the engineering and the production departments.

Computer systems involve the production of software as well as hardware. Because of the unique characteristics of a computer program and the specialized disciplines and types of supervision required for its creation, project managers must adopt special techniques and procedures for the software portion of the project. In addition, a close liaison must be maintained between the hardware engineers and the software computer scientists as the two products are created to make sure that they function together as an integrated entity. For hardware, the engineering department is primarily concerned that the production methods do not adversely affect the fidelity of the system or component operation as designed. In addition, engineering personnel must provide consultation regarding production problems that arise which might affect the equipment design. For instance, a soldering process might expose a particularly delicate component to excessive heat, thereby degrading its function. The design engineer would be called upon to substitute another component or to work out some procedure with the production department whereby the production difficulty could be overcome.

For software, the production or program writing, which includes coding, is closely associated with the design phase, since the same personnel often implement both types of effort. Thus the line of distinction between design and production is less pronounced for software than that which usually exists for hardware systems. Difficulties experienced with early efforts involving the creation of software have dictated that comprehensive and detailed documentation through all phases of the effort, including design and production, be maintained to accurately reflect the product.

The project manager, while concerned with problems affecting the ability of hardware and software to function as designed, is also interested in the status of the production schedule and the accumulation of costs. Thus reporting of procedures and control techniques must be implemented to guarantee to the maximum degree that the objectives of the contract schedule, cost allocations, and technical requirements of the equipment will be met.

17.2   Production Planning and Control of Hardware

In the line organization, the planning and controls for the production of the hardware are provided by the project manager in conjunction with the head of the manufacturing department, who is delegated the authority and responsibility for carrying out the effort with the resources allocated to the project. In the matrix, a three-point dialogue involving the project and functional managers and the discipline supervisor for production takes place to establish the planning and control for the production equipment.

The planning effort involves the coordination and scheduling of equipment, personnel, and materials required to accomplish the task as scheduled. The control effort involves setting the plans into motion by releasing the orders and monitoring, inspecting, and recording progress so that a continuous comparison between the planned and actual results can be made. Since project managers are responsible for meeting the schedule, budget, and specification objectives, they must work closely with project production heads (the functional manager and the discipline supervisor in the matrix). They must also ascertain that the methods and means of control are carried out so that the production effort will result in meeting the various goals.

The manufacture of the first article, or a prototype piece of equipment which is the result of a development and engineering effort, involves special planning. One unique feature is that generally not all the detailed information and drawings will be available at one specific time. The program schedule, which is usually critical, often requires that manufacture be accomplished in phases as the required information and drawings are released for production.

Because of the piecemeal nature of the manufacture of prototype equipment, the planning and control must be supported by some logical and feasible division of the equipment elements. The division of the production effort into elements is analogous to the work package concept of the PERT plan. The work package generally relates to the circuits and elements contained in a chassis (when dealing with electronic equipment). However, with the use of integrated circuits and other highly compact modules, a single chassis can accommodate several different systems, each of which is described in a separate work package. When a simple chassis includes modules described in more than one work package, the production plans must provide for such a combination to be performed as a single task.

Project production heads must establish a phasing chart for the different areas of work under their cognizance. The start of production in each area is contingent on receipt of drawings from the drafting section of the engineering department, which in turn is contingent upon the completion date of the design effort. There are certain exceptions to this procedure related to the production of the different standard components and modules; these procedural exceptions are established by the project manager and the project design engineer. Standard components—such as connectors, sheet metal work for consoles, standard servo units, and similar items—would be fabricated, purchased, and stored for use at future points in the program.

The creation of a phasing schedule is an effort coordinated among the various departments involved. In this case, the individuals involved are the systems engineer, the project design engineer, the project production coordinator, and the schedule and cost coordinator, all of whose contributions and inputs are coordinated by the project manager and, in the matrix, by the functional manager.

One point of interest that should be noted relates to the advantage of having the drafting section under the cognizance of the project design engineer. Any slippage in the design effort and/or drafting effort is the responsibility of one individual, who can apply whatever accelerated effort may be required to get the release from drafting on time and thereby avoid delays in starting the production cycle.

As far as the production effort is concerned, the initial concern of the project manager is that the drawings in the various areas be completed and released to the project production head as scheduled. To effect adequate preparation, the project manager receives periodic schedule prediction reports from the planning and control head that are derived from information compiled by the project design engineer. The reports, similar to the report in Figure 11.6, give information regarding each of the areas of the phasing chart. They alert the project manager to any difficulties that exist and permit remedial action to be taken as early as possible.

It should be noted that any information regarding the schedule and cost of the project is received and correlated by the planning and control head prior to being forwarded to the project manager. There are many advantages to such a procedure, especially if the company has more than one program under contract. Planning and control heads, having the overall picture of the operation of the company, can present an evaluated and objective report of the status of the program. They are also able to relieve the various department heads and the project manager of the details of putting the information into the necessary form.

Once the production effort is initiated in any area, the project manager is primarily interested in whether the schedule is being maintained, whether the accrued costs are within the budget, what courses of action are available if either the budget or the schedule is not being met, and what effects any possible design change might have on the cost and delivery.

Project managers derive their basic information from three reports similar to those discussed in Chapter 11: the cost prediction report (Figure 11.6), the schedule prediction report (Figure 11.7), and the line of balance report (Figure 11.8). The information for the reports is accumulated by the project production head by means of the cost-accounting system adopted for the project and is correlated by the planning and control head for presentation to the project manager.

In establishing the schedule status of the program at any particular time, consideration must be given to the characteristic schedule curve of accomplishment. This graphic presentation is referred to as the S curve.

Figure 17.1 illustrates the S curves for the production and assembly schedule of the reflectivity and the elevation processor units of the data processor sub-

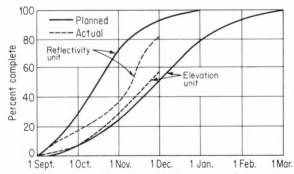

FIG. 17.1 Characteristic schedule curve of accomplishment (S curves) for reflectivity and elevation units of the data processing subsystem.

system of the RLM simulator. The solid lines show the planned schedule for each system. The dotted lines represent the actual progress of the production of each system.

Reference to Figure 17.1 will reveal that as of December 1, the reflectivity unit was 82 percent completed, whereas it should have been 95 percent complete. The elevation system, on the other hand, is shown to be ahead of schedule, since as of December 1 it was 58 percent complete as opposed to the planned completion of about 50 percent.

One important characteristic of the S curve that should be noted is its shape. At the start of the effort, the slope is almost horizontal, increasing rapidly to become almost vertical; it then decreases to become horizontal again. The shape reflects the fact that at the start of the work, relatively little is accomplished because of the necessary planning. Progress is made most rapidly during the middle portion of the period and then slopes off again to the finishing-up of the effort.

The slope of the elevation system production S curve indicates that the work requires a large amount of planning and preparation so that its slope will be relatively shallow during the initial phase of the effort as compared to the detector system S curve.

The prediction reports are derived from the S curves. Figure 17.2 illustrates the production prediction schedule report for the reflectivity unit. In comparing Figures 17.1 and 17.2, the correlation of the slippage in schedule can be noted, and the indication of the results of the corrective action is shown by the reversal of the slope of the curve in Figure 17.2. It should also be noted that the corrective action was not taken soon enough; nor was it adequate to permit the scheduled completion date of January 1 to be met.

The project manager in the line organization uses the information presented in the forms discussed earlier to report to higher authority and take other, more drastic corrective action as necessary. In the matrix, to secure additional resources, the project manager discusses the problem with the functional manager.

The same type of reporting medium is used for the comparison of actual costs with budgeted figures. The cost prediction report is prepared for the project manager so that the trend of production costs in various areas can be observed and so that the necessary corrective actions may be taken if the cost experiences are unfavorable. The possible courses of corrective action that would be taken in the event that the cost or schedule picture were unfavorable or if there were indications that it would become unfavorable would depend on the analysis of the situation and the conditions that prevail. The main point that the project manager should recognize is that when there are indications that there will be a slippage in the schedule or that the cost will exceed the budgeted costs, corrective action is required as early as possible.

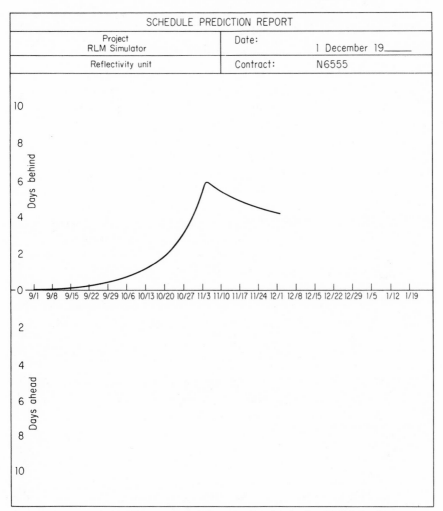

SCHEDULE PREDICTION REPORT		
Project RLM Simulator	Date:	1 December 19____
Reflectivity unit	Contract:	N6555

FIG. 17.2 Prediction report for production of reflectivity unit of RLM simulator.

17.3 Production of Software

The creation of software proceeds in a logical sequence of phases which are depicted in Figure 13.1. Program writing comprises organizing the different routines, performing checkouts, and documentation as well as the coding effort itself. In the more restricted sense, coding is generally considered to be software production and analogous to the production phase of hardware.

Thus in the coding process the computer program design is converted into a special coded language that the computer can execute. The coding effort involves a step-by-step translation of the software design in the same sense that the design is a translation of the functional requirements specified for the system under development. The primary characteristics of the program code are that it must be an accurate translation of the program design, it must be reliable, and it must be maintainable and capable of being traced, corrected, and revised. To achieve these major qualities, the project manager must make sure that the configuration management functions depicted in Figure 14.5 are effectively executed. The program design detail (PDD) document must be accurate and comprehensive, since it is the specification the coder utilizes. Verification of the acceptability of the PDD is usually established as one of the objectives of the critical design review in which the project manager and the programming chief participate. In addition, the program manager must make sure that during the coding adequate testing of single and routine, as well as multiple and related, functions is implemented for verification of accuracies and fidelity.

The identity of the programming language to be used is either specified as in the case of the RLM simulator or established during the earlier analysis and design effort. Most developing systems adopt high-order languages, since their use facilitates the coding effort and they are easier to read and maintain than the more specialized assembly and machine languages. The project manager should avoid deviating from the use of high-order language and should permit coding in assembly language only when use of the high-order language is not feasible.

The coding effort is greatly facilitated if the computer hardware is available for checking the work as it progresses. In addition, the availability and utilization of software tools are essential to the efficient achievement of the quality objectives for the software.

In the line organization, it is essential that the project manager discuss prerequisite tooling and resource requirements with the programming head and issue formal directions. A similar dialogue between the functional manager and the discipline supervisor heading the programming effort is necessary in the matrix to establish the resource requirements and the scheduled use of the resources in the coding effort.

To ensure the production of a software product of an acceptable quality, the project manager must also establish and issue applicable standards and guidelines. Such coding directives are established after discussions with the programming head. Although the program manager may not be a qualified programmer, an understanding and an appreciation of some of the fundamentals of the art are required. Examples of common standards that might be applicable to the project coding effort are as follows.

1. Supplement the code with comments on what the routine does and how it accomplishes its function.

2. Maintain simplicity in the code and avoid instituting trick codes in attempts to save memory or time.

3. Avoid special language features.

4. Names assigned to variables must be meaningful and consistent.

5. Implement the most direct and simple code structure at all times.

6. Utilize single-entry and single-exit code structures for each module.

During the coding procedure, verification tests must be performed on both single routines and routines comprising multiple functions that are interrelated. The tests for single functions are sometimes referred to as thread testing, and the tests for multiple functions are sometimes called build testing. Regression testing is another type of verification procedure used to identify any adverse effects that a newly completed code might have on existing coded routines. It is the responsibility of the project manager to make sure that plans and procedures for verification testing of the code as it is completed are established and rigorously implemented, that each test is completely and accurately documented, and that the resulting documents are filed for review as required. The tests performed during coding are preliminary supplements to the more formal system, integration, and acceptance testing that takes place when the computer program and hardware systems are complete. The tests performed during coding are vital, since errors are at that time far more easily identified and corrected than they are during the later system tests.

In the discussion of software design in Chapter 13, it was noted that a design walkthrough is implemented prior to initiating the coding effort in order to verify the complete and accurate transformation of the specification into the software design. When the coding is complete, a code walkthrough is required to verify the transformation of the software design into the code.

The procedures for the code walkthrough vary for different projects, but the common objectives are as follows:

1. Identify and correct errors in the code.

2. Identify inconsistencies between the code and the design.

3. Establish the completion of the coding effort and initiate the formal test phase.

17.4 Software Tools

The creation of computer software involves a specialized, time-consuming effort that is often described as tedious. Because of the increasing complexity of computer systems and the improved ability of programming techniques, software costs represent the major expenditure for many computer systems, and the trend continues in that direction.

To facilitate production and testing and to realize economies in developing software, a variety of software tools has been and continues to be developed for use with computer projects. Software tools include computer programs that aid and automate the creation of computer programs. It is the responsibility of the project manager to promote the utilization of the largest possible number of software tools applicable to the project.

Some common software tools and their functions are as follows:

1. *Compilers:*   Used to translate high-level programs into machine codes that can be executed
2. *Editors:*   Automatically provide for revision of programs without coding forms or format changes
3. *Analyzers:*   Collect data from the source statements of other programs by means of scanning techniques
4. *Program library:*   Stores and retrieves program modules that may be required during program writing

A computer program software tool is written in a specific language for a particular function. However, when similar computer programs are involved the tools can be used or modified for use on other projects in the company or made available to other organizations. As a result, libraries are maintained from which particular software tools can be made available for use on a particular project. A project manager should direct that the programming chief maintain a library of programming tools and utilize them to the maximum extent.

In addition to special computer programs, documents (such as specifications and test procedures) and such procedures as walkthroughs are generally considered to be programming tools.

17.5   Hardware Quality Control

On a project similar to the RLM simulator, quality control in production embraces two general areas. The areas in question are the module assembly and the quality of various components.

The objectives of reliability excellence are synonymous in many respects to those of quality, so that the achievement of reliability often automatically results in the achievement of quality. The excellence of the final product depends to a large degree on how well reliability is incorporated in the design and how well quality is achieved in the manufacturing and assembly process.

The factors that constitute quality of a product differ for different types of equipment. For instance, an electronic system operating at low frequencies can be built to meet satisfactory quality standards with little or no attention given to shielding, length of leads, etc. However, for a high-frequency system, these factors are very important, and failure to exercise the proper control in production in these areas will result in a product that, as far as quality is concerned, is unacceptable. The quality control engineer must therefore exercise proper controls to guarantee that the production department recognizes all problems and that it achieves the quality objectives that have been specified. It is particularly important to the project manager that the project production head implement a program that will achieve the required quality objectives.

Another area of quality that concerns the project manager is component quality and the means by which it is achieved. Quality control, in the traditional sense, involves the statistical sampling of a component as it is produced or delivered by a supplier.

When a volume of a component is produced or procured, it is often impractical to check every unit to determine whether the components meet the specification requirements. Therefore, a calculated risk must be assumed by the user as to whether a particular batch meets the specification requirements.

The normal distribution curve discussed in Figure 11.4 would be used as the basis for the statistical sampling of a production run of a component. The tolerance that production coordinators are willing to accept and the risk they are willing to assume would be correlated with the standard deviations of the curve to meet the desired sampling criteria.

Integrated circuits possess an inherently high quality, particularly in the area of reliability. Quality standards for integrated circuits which perform different functions have been established for various government agencies and other organizations. Therefore, the selection of items from listings of approved integrated circuits performing the desired functions will, in most cases, provide a significant degree of assurance that the desired quality of performance certified for the items will be realized in the system.

In addition, the unit and system tests will verify the performance objectives or reveal any substandard units that may have escaped the parts screening process.

17.6 Software Quality Control

The achievement of quality in both hardware and software requires that quality features be considered in the design as well as in the production phases. In creating hardware, the emphasis on quality takes place during the production phase, when physical properties such as component quality, workmanship, tolerance limits, etc., are involved.

Since software production does not involve physical properties that would affect its quality in any significant way, the quality of software is primarily determined by its ability to perform the functions for which it was designed.

The quality of software is determined by the accuracy and completeness of the documentation of the design, the code, the interfaces, and other features of its design. The excellence of the documentation will, among other things, determine the traceability of the software design, which determines how effectively the program can be reviewed and verified.

The achievement of quality assurance must be considered with equal emphasis during the analysis, design, test, and the coding phases.

The procedures for achieving the desired software quality vary among organizations. However, since the government is the major user of software, many documents relating to software development and quality assurance have been published. Most procedures and criteria used by various government activities and industrial organizations are based on government practices and experience.

The fundamental procedures for achieving software quality assurance include the following:

1. Provide detailed and comprehensive documentation as the software is developed.
2. Conduct frequent reviews of the product in accordance with an established milestone schedule.
3. Conduct tests of routines and systems to verify that the quality objectives are being achieved as the software is developed.

In the final analysis, the goal of the effort that the project manager directs is to achieve a computer program which includes the following characteristics of high-quality software:

1. *Reliability:*   Ability to perform all specification functions without significant errors
2. *Usability:*   Ability to utilize, maintain, revise, etc., without special procedures being required
3. *Maintainability:*   The degree to which structure, design, and documentation of software permit ease of maintenance and modification
4. *Validity:*   The degree to which the program provides for all the functions necessary to satisfy the requirements and for effective interfacing with associated software elements
5. *Testability:*   The degree of use of simple structure, general algorithms, and logical design that enhances the testing of all functions
6. *Traceability:*   Ability to correlate each element of a software with its associated documentation
7. *Robustness:*   Ability of the program to recover from erroneous or inconsistent inputs and continue without detrimental results.

17.7   Summary

The production plans and techniques vary for different types of equipment. For a development prototype piece of equipment, the production plan would be in essence a custom-built device.

Project managers are primarily interested in the cost and schedule status of the production effort and implement a reporting system by which such information can be transmitted to them.

The development of a computer system involves software as well as hardware. Because of the unique characteristics of software as distinguished from hardware, special management procedures must be implemented to make sure that the computer program will possess acceptable quality and functional characteristics.

For hardware, tracking of production status is accomplished with a variety of documents, such as the line of balance and cost prediction reports. In addition, quality control is achieved by those production control procedures which will provide the greatest degree of assurance that the quality and reliability design features are implemented. Because of the inherently higher reliability that integrated circuits possess as compared to earlier design involving transistors and similar components, quality and reliability objectives are achieved primarily by selecting those standard integrated circuits that have been included on the approved list of government and industrial organizations and which possess functional characteristics which satisfy the design objectives.

The coding phase of software is akin to the manufacturing phase of hardware, although the software design and coding phases are intimately interrelated. To achieve a successful coding effort, it is necessary that standards be established by the project manager and then strictly followed. To ensure an acceptable quality of software, complete and detailed documentation of the coding effort and tests is required. Upon completion of the coding effort and prior to conducting formal unit, system, and integration tests, a code walkthrough is required to verify the acceptability of the computer program.

BIBLIOGRAPHY

Driscall, A. J.: "Software Visibility and the Program Manager," *Defense Systems Management Review*, Washington, D.C., vol. 1, no. 2.

Fife, D. W.: *Computer Software Management*, NBS Special Publication 500-11, U.S. Department of Commerce, Washington, D.C., 1977.

Houghton, R. C.: *Computer Science and Technology*, NBS Publication 500-74, U.S. Department of Commerce, 1981.

18

TEST AND
CHECKOUT

18.1 Test Objectives

The creation of any product requires that some procedure be adopted and followed for establishing whether the product is consistent with what was intended. For nondynamic items such as shoes or food products, the tests would represent quality control checks relating to tolerances, grade of materials or ingredients, workmanship, etc. For a piece of equipment which is designed to perform in a specified manner, the tests would cover a much larger area and would necessitate the measurement of dynamic results. The product manager who is involved with dynamic equipment is responsible for the type of tests that are necessary to establish whether the equipment being produced satisfies the requirements specified by the contract.

The test and checkout of equipment being produced as a development effort or a new design are a continuing effort throughout the program. The testing falls into two basic categories: verification tests, which the contractor performs to be sure that the equipment will function as specified, and the customer's acceptance tests, which determine whether the equipment meets the requirements of the contract and specification.

For computer systems, a different sequence and different types of tests are required for hardware than for software. When the tests on each of the products of the computer system have been successfully completed, integrated tests are performed to determine how well the complete system satisfies the originally established performance objectives. One of the primary purposes of the separate testing of hardware and software (particularly the software) is to identify and correct any errors, omissions, etc., as early as possible. Since corrective actions that must be made during the integrated test phase are extremely expensive and time-consuming, a comprehensive and detailed test and verification effort during the design and production phases for the hardware and particularly for the software will minimize problems during the final integrated and acceptance testing of the system.

The types of hardware tests performed by the contractor would include the following:

1. *Breadboards:* To verify the merits of a particular circuit or module design

2. *Subsystem:* To establish the performance and compatibility of specific subsystems of the equipment

3. *System:* For verification of overall equipment performance

4. *Reliability:* To determine ability of elements, modules, and circuits to perform within acceptable reliability limits

5. *Quality and assurance:* To verify acceptability of components, workmanship, arrangements, etc.

During software development, the following types of software tests would be performed.

1. *Module or unit tests:* To verify the lowest software element that represents a functional entity

2. *Multi-unit tests:* To verify that module groups are interfaced in such a way as to to provide required results

3. *Function or system tests:* To validate the acceptable performance of a system or subsystem

4. *Integration tests:* To confirm that the computer program can execute its functions, through the interfaces of the system, in an acceptable manner

In addition to the different types of scheduled testing, verification (which includes reviews, audits, and walkthroughs) is required throughout the development effort to make sure the performance requirements of various modules of the program are being met and that documentation is complete and accurate. The verification effort, which is required throughout the software development cycle, is generally considered to be an essential part of software testing.

The customer's primary concern is to determine whether the equipment specified in the contract satisfies the contract requirements. The tests performed under the jurisdiction of the customer include system tests for hardware, system integrated software and hardware tests for computer systems, and reliability, maintainability, and environmental tests. In essence, the purpose of the customer's acceptance tests is to establish that the equipment completely conforms with the terms and requirements set forth in the performance contract specification.

18.2 Hardware Test Criteria

For a piece of equipment that is being engineered in response to a performance specification, the detailed test criteria must generally evolve with the design effort. The performance specification may provide equipment objectives, such as general accuracy criteria, response items, etc., but the detailed characteristic performance curves, varieties of functions, and many combinations of performance modes are not generally expressed in the performance specification. Because of the nature of such projects, the test criteria are developed during the design phase and represent a mutually agreed-upon vehicle for use in establishing the acceptability of the equipment.

The performance specification also identifies the source of data that is to be used in designing the equipment. Generally, the detailed data supplements the performance specification, and the acquisition and use of the data are responsibilities that the contractor assumed in accepting the contract. For example, the detailed data relating to the radar antenna beam pattern in the RLM simulator must be procured by Creative Electronics in order to simulate the radar returns in the manner required by the specification.

In developing a hardware system, each module or subsystem is designed to achieve specific performance objectives which contribute toward achievement of the system performance requirements of the specification. To make sure that the performance of a particular module satisfies its individual criteria, the engineer must establish performance and design criteria and identify the type of tests to be used to verify that the criteria will be met by the module in question. The specific criteria and the identity of the tests would be documented as a facet of the design effort. Furthermore, the results of any such tests performed during the design phase would be documented for future review and reference.

Similar procedures for establishing the criteria and tests for subsystems and ultimately for the total system would be required in order to verify that the performance of the system provides the results established by the specification.

18.3 Software Test Criteria and Procedures

The testing of the software follows the same general pattern used in the testing of the hardware, except that the test effort is more intimately associated with the design effort for the software. Since one of the desired qualities of a computer program is ability to be easily tested, the features of test criteria and procedures are considered in all of the phases of the software development cycle. In many cases, an independent test team headed by a test manager is designated as part of the project team reporting directly to the project manager. Concurrent with the system engineering effort, a review of the proposal, the contract, and other pertinent project documents is made in conjunction with the preparation of a test plan to be used for the software. The plan, which is coordinated with the project manager, the user, and other interested parties, is used as a guide for verifying the software during its development cycle.

In designing and coding modules, module groups, and systems, the programmers identify the various desired output functions which require verification. The test criteria and procedures for the different levels of the software product include the objectives of the different reviews that take place, such as the design walkthroughs, the critical design review, the code walkthroughs, and other management reviews and audits of the development effort.

Examples of the different categories of errors that are uncovered during testing are computer program deficiencies, document errors, design flaws, and logic difficulties. Each of the errors is prioritized according to the severity of the effect of the deficiency upon the operation of the system. Based upon the category of error and its priority, the timing of corrective action and the comprehensiveness of rechecking by the contractor are established. Further, the number of types of computer program patches that are required as the corrective actions are established as one of the elements of the checkout procedures. The foregoing are some of the points relating to the software checkout that must be agreed upon by the project manager and the representatives of the customer.

The effectiveness of the debugging and testing effort of software is contingent upon scrupulously correct and complete documentation. When satisfactory documents are available, different checking aids are employed for the testing at all levels. Some of the aids with which the project manager should be familiar are as follows.

1. Compare programs to identify logic errors.
2. Trace programs to locate instruction errors.
3. Memory drawings to produce specific listings of memory information.
4. Trap programs to select computer information for review.

The results of acceptance tests have contractual implications involving significant sums of money. Therefore, each party to the contract—the user and the developer—has a vested interest in the identity of the test criteria and the results of the tests. The contractor or developer of the equipment does not want any performance criteria included in the test procedure which exceed the performance requirements of the specification. On the other hand, the user wants to make sure that all significant characteristics and criteria included in the specification are adequately covered in the test procedures for the equipment. The identity of the criteria and procedures that are included in the test procedures document is discussed, negotiated, and ultimately established by mutual agreement between the contractor and the user. The project manager has the prime responsibility for assuming that the criteria and tests agreed upon are consistent with the contract terms.

18.4   Project Testing Guidelines

The test program to be used for any system has to be tailored to that system on the basis of factors such as size, complexity, stringency of performance requirements, and type of system design. In order to provide adequate testing of any system, basic parameters should be included in the test plan. An effective test plan for computer systems would include the following:

1. Independent test team under the direction of test supervisor reporting directly to the project manager (and to the functional manager for matrix organizations)
2. Schedule of required resources, such as personnel, tools, and equipment, to execute the testing effort.
3. Criteria and procedures to be used to effectively test modules, systems, and integrated hardware-software outputs
4. Assurance that the hardware circuits and software modules are designed to facilitate testing at all system levels
5. Implementation of top-down integration testing procedures wherever possible
6. Rigorous application of configuration management disciplines to the testing program
7. Complete and accurate documentation of all procedures and results.

Although the conditions cited above relate primarily to computer systems, many of them are applicable to projects that involve only hardware and should be applied as appropriate.

18.5   System Integration and Acceptance Tests

The testing of the integrated hardware-software portions of a computer system represents the key indication of how well the system was produced. Results of integrated testing of a complex system are rarely if ever acceptable during the testing's initial cycle. The magnitude of corrective actions generally is a function of the number of improper actions taken during previous phases of the project cycle. Some major types of mistakes that have caused problems during integration and acceptance testing are as follows.

1. *Uncoordinated integration effort:*   An orderly and consistent plan for integrating the various software and hardware elements and modules of the

system was not implemented. The failure to rigorously control the step-by-step procedures relating to the design, production, and testing of the equipment subsystems usually results in incompatibility of the software with the hardware elements, which requires time-consuming and expensive corrective actions during integration testing.

2. *Deficient plans:* The failure to provide detailed procedures for integrating hardware and software leaves aspects of the effort open to interpretation. Loopholes in such plans will eventually result in some action that will have an unsatisfactory effect. A deficient plan also includes documented procedural errors which reap unacceptable results.

3. *Inadequate documentation:* During the integration effort, which is often conducted in an environment of pressure and haste, changes and modifications are made without the proper follow-up of detailed documentation. Other memebers of the integration team, checking the same or related functions, experience difficulty and frustration in trying to identify and correct the problem as a result of receiving what then constitutes incorrect data.

To minimize integration difficulties, it is incumbent upon the project manager to strongly enforce project directives such as the requirement for comprehensive documentation during the integration tests.

Subsequent to the satisfactory completion of the integration tests, necessary document revisions are made to reflect actions required by the tests and the acceptance testing of the system is conducted. The results of the acceptance tests are reflected in the final test report, and decisions are made as to whether the system should be turned over to the user or further corrective actions should be taken before transfer of title.

18.6 Scheduling of Test and Checkout

The test and checkout of the finished product form what is in many respects the most critical effort, especially in relation to the schedule. Because the integrated system test and checkout always occur at the end of the program, there is very little, if any, slack time in which any unforeseen difficulties can be corrected. A further impediment to swift corrective action is the fact that the test and checkout occur when the system is packaged in its final configuration. As a result, troubleshooting, taking corrective measures, or incorporating modifications constitutes a difficult and time-consuming task. When difficulties of even a relatively minor nature occur during the testing and checkout of the integrated system, the project invariably slips to the point at which delivery of the equipment in accordance with the schedule is impossible.

It is therefore important that detailed plans be made for the test and checkout effort. For a prototype system which involves development, the engineering department would perform the tests. The schedule and plans for such a test program would be made by the project design engineer, the programming chief, the project manager, and the schedule coordinator, with contributions by the various engineering group leaders who would be doing the actual work.

18.7 Summary

The testing and checkout function is a continuing function throughout the cycle of design fabrication and performance verification. As the design is created and corrected, the test criteria for the product is developed. The end result is that there must be consistency among the equipment performance, the data and requirements upon which the design is based, and the criteria for the tests.

Computer systems require that a test program be implemented for the software that is under development as well as the hardware. In addition, the computer system test program must provide for verification of the integrated functions of the hardware and software.

The project manager is responsible for establishing a test plan which includes an independent team to conduct the tests, scheduling of required testing resources, comprehensive documentation of the testing, and, in general, assurance that the tests that verify the acceptability of the system are conducted in an efficient and thorough manner.

BIBLIOGRAPHY

Fife, D. W.: *Computer Science and Technology*, NBS Publication 500-11, U.S. Department of Commerce, Washington, D.C., 1977.

Fischer, K. F., and M. G. Walker: *Digital Systems Development Methodology*, Computer Sciences Corporation, Falls Church, Va., 1978.

Military Standard Trainer: System Software Development, MIL-STD-1644, U.S. Naval Training Equipment Center, Orlando, Fla., 1979.

Planning and Control Techniques, AMCR-11-66, Headquarters, U.S. Army Materiel Command, Washington, D.C., 1963.

MONITORING
AND SUPPORT
SIDE ITEMS

19.1 Requirements for Side Items

During the life cycle of a development project, personnel of the using activity require different types of engineering and status reports to establish how well the contractor is progressing toward meeting the technical, cost, and schedule objectives of the contract. In addition, the user is concerned about whether the documentation, spare parts, special tools, and similar items will be adequate for the effective operation, maintenance, and modification of the equipment which will be delivered.

To provide for items that will permit the monitoring of the development effort and support of the equipment after delivery, the contract usually requires the delivery of various items to fill those needs. The supporting items are identified with and necessary to the installation, maintenance, modification, and operation of the major items under procurement. Examples of such items are manuals for the equipment, spare parts, engineering drawings, special tools and test equipment, software tools, and courses of instruction for the customer's maintenance and operating personnel.

In addition to the support items, the contract will require delivery of specific engineering and status reports, technical liaison services, and similar items deemed necessary by the customer in order to effectively monitor the technical progress, schedule, and, where applicable, the cost status of the project.

The various support and monitoring elements are usually referred to as "contract side items"; the term encompasses all deliverable contract items with the exception of the main equipment.

All too often side items are given only minimal attention, with the result that the contractor, in underestimating the effort required to deliver the side items, can suffer financial losses on a contract, even though the main contract item may have been successfully completed and delivered. The difficulties noted above are not surprising when one considers that the cost of providing the side items is usually a significant percentage of the total contract selling price.

The hardware side items which are usually the cause of any difficulties are the engineering design reports, installation and maintenance manuals, and drawings. The software items which require special attention are the various tools for testing and the coding documents which must be created and provided for revising and correcting the computer program.

The specification and contract requirements governing the format and content of the reports and manuals are usually very explicit and detailed. Contractors who have experienced losses and/or damaged performance reputations due to these side items can generally point to the following reasons for their experiences:

1. Failure of project personnel or their staffs to study or comprehend adequately the detailed requirements of the side items.

2. Failure of project supervisors to assign qualified personnel to producing each of the side items or to subcontract those side items if the contractor has neither the facilities nor the personnel necessary for the completion of the item.

3. Underestimatation of the scope or complexity of the required tasks. Because of this, the project supervisor may assign such tasks as collateral duties to existing personnel who may not be qualified or motivated to complete the task in a successful manner.

19.2   Monitoring of Hardware

The monitoring of the technical and delivery status of a project is accomplished by means of the following types of documents and procedures.

Engineering Reports

Engineering reports constitute a category of technical monitoring items and logically should be produced by the contractor's design engineers according to their respective areas of design responsibility. However, the actual task of writing the report might be more expeditiously performed by technical writers working in close liaison with design engineers. Technical writers are normally associated with a special technical writer group responsible to the project design engineers, from whom they would receive their writing assignment, schedule, and directions.

Prior to initiating their work in planning a particular engineering report, technical writers must study the specifications and contract schedule to establish what areas of the report must be emphasized; what degree of detail should be used in the report to describe the subject matter; what format and arrangement of information should be used; to what extent sketches, block diagrams, and sample calculations must be used; and other similar information.

In addition to the content of the report, technical writers must be cognizant of the contract submission date of the report and the number of hours that have been allocated in the program to completion of the document.

Upon receipt of their assignment, technical writers must study the proposal used to obtain the award of contract and find out whether any changes in design approach or content have been adopted since the submission of the proposal. After doing the preliminary study, technical writers follow the organizational procedures for drafting an outline and use the information provided by the design engineer for writing the engineering report.

After the completion of the final draft, the report is transmitted to the design engineer, project design engineer, and project manager for approval prior to formal printing and reproduction.

The approval of an engineering report generally triggers a design freeze for a particular item of equipment, and the entire project schedule is often based upon establishing the design freeze by a certain date.

Since the design freeze date is a critical date upon which the schedule for the subsystem covered by the particular report depends, it is essential that the content of the report be in accordance with the relevant contract requirements. Disapproval of a report requires that the contractor spend the time to implement corrective action, which may extend the approval dates of the resubmitted documents beyond the original design freeze date. Delays in securing approval of engineering reports often require that the contractor establish new freeze dates, which in turn could require revision of the entire project schedule.

Status Reports

A variety of reports addressing the status of delivery and cost expenditures (for other than firm fixed-price contracts) are included as deliverable side items of a contract. The type of status report depends upon the cost and schedule tracking system agreed upon by the contractor and user or by what the contract specifies. PERT, line of balance, configuration management, and prediction development reports are typical. On large complex systems, the status reporting systems are often computerized and the reports are submitted in the form of computer runs.

Meetings and Reviews

In most contracts, the general terms allow for user representatives to review the status of the contract as it unfolds. Such reviews include visits to the contractor facilities, discussions with personnel assigned to the project, and informal meetings. Since formal reviews and meetings involve the expenditure of contractor resources for the preparation of agendas and presentations and other time-consuming efforts, contracts usually include reimbursable line items to cover the costs of meetings designed to explore and establish for the user the technical, schedule, and cost status of the project. Meetings may be related to design freezes, configuration management audits, test procedure discussions, and/or other subjects that are deemed necessary and are identified in the contract schedule.

19.3   Monitoring of Software

The following are some of the major items and procedures used for the monitoring of the status of software.

Development Reports

These documents are comparable to the engineering reports for the hardware systems. Separate reports address different areas related to development of the computer program, each of which is essential to the success of the product. Among the reports which would be submitted for review and approval by the monitoring personnel of the user are the program design document, the interface design specification, and the data base design document. These three reports are described in Table 13.1 and are typical of the contract side items which the project manager must provide for and deliver to the user as contract items.

Status Reports

Unlike hardware, software has no physical characteristics by which progress can be estimated. Because of past difficulties in monitoring the status of software development, one of the prime characteristics that management has advocated for software is visibility. Providing a plan for developing software that can be monitored to evaluate its design, coding, testing, cost, and schedule status satisfies the visibility requirement.

Two prime side items which require the rendering of status reports are the software development plan and the configuration management plan.

The software development plan provides a description of the software design and includes the milestones and schedules for the effort. Thus the using activity can track the software effort by securing information on the status of specific software areas and comparing this information with the information in the software development plan.

The primary monitoring information that the software configuration management plan provides is when all the requirements that establish a baseline are met. The different reports discussed in Chapter 14 indicate the stage of development of the computer program and therefore provide status information of interest to the user.

Meetings and Reviews

During the development of a software project, periodic audits and walk-throughs of the software design and code are implemented to establish the status and validity of the work accomplished. In addition, formal review meetings are held at specific milestone points to verify the status of the programming effort. The meetings include reviews of the performance specification, design specification, program description document, final software design, test plans, and other items. As noted for the hardware, formal reviews are usually identified as deliverable side items in the contract.

Other publications are the operator's and installation manuals, each specified to permit personnel to perform their specific tasks. The comments made in conjunction with the maintenance manual would also be applicable here.

19.4 Hardware Support Items

The preparation, writing, and publishing of manuals for delivery as items included in the contract represent one of the largest single side items of the contract for hardware. For a complex piece of equipment, the maintenance manual is a large volume. The amount of detail, schematics, etc., is dictated by the applicable specifications on the format and the degree of detail as well as other requirements. In general, the maintenance manual must contain the degree of detail and the supplementary documents (such as sketches) which will permit average maintenance personnel knowledgeable in the technical area represented by the equipment to be able to troubleshoot, repair, and maintain the equipment by referring to the manual alone.

The writing and printing of a maintenance manual are specialized tasks, and except in very large organizations that may have their own personnel and facilities for the task, maintenance manuals are subcontracted to publication firms.

Regardless of whether the task of producing maintenance manuals is accomplished in-house or subcontracted, the design engineers must provide the required technical information to the technical writers and maintain a close liaison with the writers.

Those undertaking the writing of the manual must be provided with the specification requirements, the schedule for submission of the different drafts, and all other pertinent information. Because of the scope of the work and the amount of effort required, those undertaking the task of furnishing the maintenance manuals must carefully study the specification and contract requirements and make all possible effort to achieve approval on the first submission.

The drawings also involve significant effort and constitute a major cost factor.

The only additional point to make is that the project manager should make sure that the chief draftsman has analyzed the format and other requirements for the drawings so that there will be no rejections of the drawings as delivered on the contract due to the fact that specifications were not followed.

19.5   Software Support Items

Some of the items that are commonly specified for the support of software are as follows:

1. *User manuals:*   These provide detailed information to enable the trainer operator to utilize the computer program. They describe the capabilities of the computer program and the various functions and give other information required for the operation of the computer system.

2. *Maintenance manuals:*   These provide information on each of the subroutines of the computer program addressed. In addition to the program standards and conventions that characterize each program, the document provides environmental data and information to enable the maintenance personnel to maintain the program, implement required program revisions, and otherwise service the computer program for effective utilization.

3. *Software tools:*   These include software programs to assist in maintenance, troubleshooting, readiness verification, alignment checkout, and other functions that would be performed by maintenance personnel. Some of the software tools that would be cited as deliverable support items are compilers, debugging programs, assemblers, libraries, test programs, and linkage editors. The specific software tools identified in the contract would be those deemed necessary for support of the particular computer program under development.

4. *Software descriptions:*   These are descriptions of the structure and design of the software modules, including source codes, program listings, cross-references of listings, load maps, and other types of software data which will permit the support, revisions, and corrections that may be required.

19.6   Spare Parts

For a procurement which involves development and for which the design details have not been established at the time of the contract award, the identity, type, number, etc., of the spares to be furnished would be unknown. Therefore, that contract item would be left open as far as the specific list of spares and their cost is concerned.

After the design has been established and the parts and components to be used in the equipment have been determined, the selection of the list of spare parts for the equipment would be made as a coordinated effort by the contractor and the customer. Usually, the contractor submits a recommended spare parts list as a deliverable item, and this list is used for provision of the spares for the equipment.

The project manager's main concern as far as the spare parts are concerned is that adequate numbers of the correct components be selected as spares so that the equipment operation and utilization in the field will not be jeopardized. If the equipment must remain inoperative due to the fact that some component was not included as a spare, the general opinion of the adequacy of the equipment (or its "image") would suffer, and this would be a reflection on the manufacturer and ultimately on the project manager.

In general, the guide in choosing what parts should be provided as spares is their life expectancy. Here, the information derived from the reliability study is used as a basis for selection.

Another factor to be considered is where the equipment will be used. If the equipment will be used at some isolated base, a comprehensive spares provision will be made. The area of utilization will be converted to a time factor. In an isolated base, the time factor might be 2,000 hours of average equipment operation. At a home base where sources of parts are accessible, a time factor of 1,000 hours might be used. Thus for a component having a meantime between failures of 500 hours, twice as many units of the component would be supplied for the isolated installation. Further, the isolated installation would be provided with spares for components having expected life expectancies of between 1,000 and 2,000 hours, whereas the home base installation would not be provided with spares for such components.

19.7   Other Side Items

A contract may require the delivery of a multitude of different side items, all of which are important and require a serious effort by the contractor. Some other items which might be required on different contracts are installation services, maintenance services, and training of personnel in the maintenance and operation of the equipment.

It is important to remember that a timely and well-planned effort for providing the side items will minimize difficulties and friction with the customer during the course of the program.

19.8   Summary

The deliverable items of a contract intended to serve as tools for the monitoring of the contract and to support the basic equipment in the field are referred to as side items. These items include manuals, spare parts, drawings, software tools, status reports, and other deliverable items.

The amount of money represented by the side items is usually about 10 percent of the total contract price, which for a large procurement represents a significant amount. The contract and specification requirements for the side items are generally well defined and call for strict conformity to such details as format, content, and delivery. In their zeal to design, develop, and fabricate the major item of the contract (the hardware and software) contractors have been known to render secondary attention to the requirements of the side items. In many such cases, late delivery of the items and rejections due to failure to satisfy the contract requirements have caused a heavy penalty in unnecessary expenditure and damaged reputation.

The successful delivery of the side items of a contract requires that the project manager establish a course of action and schedule for all the items to guarantee that the specifications requirements will be met and that each item will be delivered in accordance with the contract schedule.

BIBLIOGRAPHY

Military Standard Trainer: System Software Development, MIL-STD-1644, U.S. Naval Training Equipment Center, Orlando, Fla., 1979.

FOLLOW-UP

20.1 Postacceptance Considerations

The responsibilities of the project manager do not cease when the equipment required by the contract is accepted and delivered. There are numerous areas of a contractual nature that require final settlement: residual items of the contract may still have to be delivered, and on a procurement of complex equipment the contractor may have a continuing function to maintain the equipment, provide changes desired by the customer, continue a configuration management program for the delivered articles, and render other services desired by the customer.

20.2 Finalizing the Contract

Most contracts for equipment of a high cost incorporate some form of progress payment clause wherein the contractor receives payments of up to 85 or 90 percent of the cost of the equipment item being procured. The residual money is not automatically rendered to the contractor upon acceptance. Instead, the contractor and customer review the history of the procurement, the performance of the equipment, and other terms of the contract to determine what consideration should be offered as compensation in areas of noncompliance. Some of the types of issues in which the project manager representing the customer and the contractor might become involved during the postacceptance negotiations are as follows:

1. *Equipment performance:*   The equipment may not have met every last requirement of the specification but may have been accepted because it was judged to be in "substantial compliance" with the contract requirements. In such a case, the customer, having received less than what was contracted for, is entitled to some consideration which would be equitable compensation.

2. *Latent defects:*   A latent defect of the equipment is considered to be one that became apparent only after utilization but exists due to a flaw in the design or quality of the equipment. The contractor has the obligation to remedy the defect or negotiate equitable compensation to the customer.

3. *Software deficiencies:*   Undetected deficiencies may surface in the form of computer program malfunction, inconsistencies between documents and computer programs, software design errors, and similar difficulties. Rather than assuming open responsibility to implement corrective actions subsequent to delivery, the development contractor usually provides for reimbursable on-call services or on-site support for the software by the contractor so that deficiencies during operation can be corrected. Such an arrangement usually also includes field support for the hardware. Since the support services are usually provided by a field service manager, the orderly transition of data and responsibility must be provided by the project manager.

4. *Delivery:*   Delays in delivery may have been responsible for inconvenience or expense to the customer. If the delay were the contractor's fault, compensation would be negotiated.

5. *Claims:*   Either party may have a basis for claims on the other due to some form of breach of contract. The customer, for instance, may have directed the contractor to provide some service or feature over and above the terms of the contract, and the contractor may therefore be entitled to compensation; or the customer may have failed to provide data or other items as scheduled.

6. *Rate structure:*   Since most contracts for newly developed or engineered equipment are of the incentive type, customers, particularly the U.S. government, retain the right to negotiate overhead rates, G&A rates, and other factors that make up the final contract price.

The project manager participates in the final negotiation, where each item is considered point by point and where a determination of the final price and possibly revision of the schedule might be negotiated. A question might be asked as to the logic of negotiating a final schedule after the equipment has already been delivered. When the equipment is late, contractors are interested in negotiating a schedule revision in order to have their performance record look favorable. A successful contractor must always look to the future, and one important negotiation point is the company's performance on previous contracts.

20.3 Subsidiary Items

Many of the subsidiary or side items require delivery subsequent to delivery of the primary equipment in the contract. The customer's project personnel have the responsibility for inspecting the delivered item and verifying that it meets the contract requirements. Examples of such items include the following:

1. *Drawings, software, and support documents:*   One of the disciplines of configuration management discussed in Chapter 14 is conducting a configuration audit to verify that the drawings and other material are consistent with the equipment such data is supposed to describe. The audit, sponsored by the project manager, establishes the acceptability of such material.

2. *Spare parts and support equipment:*   When such items are required by the contract, the identity, quantity, and quality must be verified prior to rendering acceptance.

3. *Maintenance services:*   The customer may require that the contractor render maintenance service on a continuous or on-call basis for a specific period of time. The project managers for both parties usually are involved in implementing the maintenance program.

20.4 Software Support

The flexibility inherent in computer systems is the result of the ease of modifying the computer program. Computer programs are usually revised and updated to satisfy changing industrial or military objectives. However, software revisions in the field must be as rigorously controlled and documented as revisions made during the development phase.

The software support includes modifications to satisfy revised requirements, correction of errors that are uncovered, creation of software for new requirements, and improvements to the computer system operation.

The software support effort can be accomplished by field service personnel, through on-call contractor service, or at a central software support facility. Software support facilities have been established by large industrial and government organizations to render support for computer systems around the world. The many advantages of such installations include efficiency of software support, the ability to provide consistent programs for similar operations, good control of documentations, effective configuration management, and the advantages of a standardized operation.

20.5 Field Reports

After the equipment is installed, the contractor will usually either be under contract to maintain the equipment or have the responsibility to correct difficulties that may occur during the warranty period, which may last between 3 months and a year. In either case, the contractor has access to the field reports describing the operation of the equipment.

The field reports are compiled periodically and describe the number of hours the equipment was utilized, the difficulties experienced, the corrective actions taken, and general comments. The project manager analyzes each report and catalogs the information to bring out any trends or patterns. For instance, if the reports indicate computer dumps or a recurring breakdown in a particular subsystem and the remedial action involves replacement of a particular component, an analysis of the software or hardware design will probably reveal a need for redesign of the subsystem. This would be implemented as a field retrofit.

20.6   Unsolicited Proposals

An unsolicited proposal can be submitted at any time, and when the U.S. government is involved, there is an obligation to review the proposal and render an evaluation. Whether anything further materializes depends on the requirement that may exist and the funds that are available.

If a company has something to offer that seems to meet a need and is worth pursuing, the groundwork should be laid by communicating with the personnel of the potential customer and by promoting the merits of the equipment to be offered.

In the case of the RLM simulator, the Creative Electronics Corporation should submit any design improvements that may correct some inherent weakness evidenced in the field service reports or which may improve the capability of the equipment. Such recommendations could be submitted as an unsolicited proposal. Close surveillance of the operation of the equipment and a close liaison with the customer will serve to maintain good customer relations and to enhance the chances of obtaining future contracts.

20.7   Summary

Project managers have a continuing role after all items of the contract are delivered. Follow-up actions include settlement of claims, renegotiation, and submission of unsolicited proposals.

For contracts which require on-call or continuous support services, the project manager is responsible for making sure that the necessary data and other information are made available to the groups that are responsible for the support of the software and hardware.

Other follow-up actions include the receipt and correlation of field reports on the delivered equipment, the maintenance of liaison with the customer, and a general effort to enhance the contractor's image in the eyes of the customer.

GLOSSARY

ACCEPTANCE TESTS: Tests to which the equipment is subjected to determine whether the equipment meets the specification requirements.

ACQUISITION PHASE: Period in configuration management plan when item is being created either by in-house or by contractor effort. Phase is divided into design/development stage and production stage which terminates at operational baseline.

ACTIVITY: An element of work effort used in conjunction with PERT.

ALLOWABLE COSTS: Costs incurred by a contractor for expenditures outside of the scope of the contract.

ARMED SERVICES PROCUREMENT REGULATIONS (ASPR): A compilation of the rules that govern the procurement of equipment and services by the U.S. government from private industry.

ASSEMBLY LANGUAGE: A computer code expressed in generalized computer language which is fed into the computer to achieve desired calculations. The computer itself will translate the assembly language into machine instructions.

AVAILABILITY FACTOR: Measure of the ability of equipment to be utilized, expressed as functions of reliability and maintainability.

BASELINE: Specific point during configuration management cycle which divides one phase from the following phase. Baselines defined as characteristic, functional, and operational terminate the concept formulation, definition, and acquisition phases, respectively.

BIT: A binary digit or a single character in a binary number. For example, 100101 (which represents the number 37) has a word length of 6 bits.

BLOCK DIAGRAM: A graphic presentation of a design in which the functions of subsystems or modules are shown as blocks and in which interconnecting lines represent the flow of signals.

BOILER PLATE CLAUSES: Standard contract clauses used in contracts by procuring organizations.

BREADBOARD: An energized interconnected assembly of components which is used to verify the design of a system.

BUDGET: Planned expenditures and commitments, by time periods.

CASH FLOW: Money coming into and out of an organization or project as income and expenditures. Net cash flow is established after charges such as depreciation, interest, etc., are considered.

CHANGES CLAUSE: Contract provisions permitting the customer (usually the government) to unilaterally revise the contract. Such changes, however, entitle the contractor to receive equitable adjustment in the contract price.

CHARACTERISTICS BASELINE: A point in the configuration management cycle which terminates the concept formulation phase and initiates the definition phase.

CLARIFICATION: Answers to questions posed by the offeror or the customer regarding ambiguities, incomplete material, etc., in specification proposals or other procurement material.

CODES: Statements that translate program design into a format that can activate a computer.

COMMERCIAL CONSIDERATIONS: Nontechnical factors which may influence a contract award, such as price, cost-sharing arrangements, and penalty/incentive issues.

COMMONALITY: A factor reflecting the percentage of components, modules, and other elements of a piece of equipment which are interchangeable with existing equipment of a different type.

COMPATIBILITY OF SYSTEMS: The ability of two or more systems to perform effectively when interconnected electrically, mechanically, hydraulically, etc.

COMPILER LANGUAGE: The computer subsystem capability of accepting high-level computer language such as Fortran and effecting the translation into assembly language or into the machine language that can be used directly by the computer.

COMPUTER MEMORY: A medium in which information relating to a computer can be stored on command and held for retrieval upon command—e.g., disc storage, core memory, etc.

COMPUTER PROGRAM: The instructions which are used to direct the operation of the computer. The instructions can be expressed in compiler, assembly, or machine language.

COMPUTER SPEED: The rate at which a computer element can function. A computer speed of 1 microsecond indicates a cycle rate of 1 million per second.

CONCEPT FORMULATION PHASE: The period in the configuration management cycle during which the desired objectives for satisfying a requirement are determined as feasible or unfeasible. The results of this phase establish the characteristics baseline.

CONFIGURATION AUDIT: Verification that product and documentation satisfy baseline and other milestone requirements.

CONFIGURATION CONTROL BOARD (CCB): Management group responsible for approving or disapproving changes in baseline documents or computer software.

CONFIGURATION ITEM (CI): Hardware or computer program contract item (CPCI) which has a specific end-use function and which is designated for configuration management.

CONFIGURATION MANAGEMENT: The implementation of formal management, technical direction, and controls during a project's life cycle to provide a complete definition of functional and physical characteristics of each item, to control the adoption of changes, and to maintain a continuous accounting of the design and equipment configuration.

CONFIGURATION REVIEW: Review to determine completion status at configuration management milestones.

CONSTRAINT: An impediment in a PERT network caused by a particular activity which prevents events from being reached or activities from being initiated.

CONSTRUCTIVE CHANGES: A communication by a representative of the procuring organization which is taken as direction by the contractor and acted upon.

CONTINGENCY FACTOR: An increment of cost or other margin to provide protection from unplanned expenditures brought about by unforeseen factors.

CONTRACT CLAUSES: Terms relating to specific rights and obligations of the parties to a contract.

CONTRACT LAW: The legal ground rules that are applicable to parties in a contract which can be enforced or resolved in the judicial system of the government.

CONTRACT SCHEDULE: A contractual document specifying deliveries, description of deliverable items, terms of payment, special contractual requirements, and other items pertinent to the contract.

CONTRACTOR CLAIM: A request for a consideration brought about by effort or expenditures for features or performance capabilities claimed as beyond the scope of the contract.

COST BREAKDOWN OR COST SEGMENTS: A segregation of total costs to identify the amounts required for specific efforts or expenditures.

COST EFFECTIVITY: The total cost sustained by a user as the result of purchasing a particular type of equipment. The total includes those costs that might be imposed by modifications or changes to existing supporting equipment necessary to utilize the newly purchased items.

COST MANAGEMENT: Procedures by which expenditures of resources and money are monitored and controlled.

COST-PLUS-FIXED-FEE (CPFF): A type of contract generally used for R&D programs by which the contractor is reimbursed for all costs incurred, but the fee or profit is a fixed amount.

COST PREDICTION REPORT: A report in which a comparison is made of the actual and budget costs at a given point in time and in which the trend of costs is indicated to facilitate cost prognostications.

COST-REIMBURSABLE CONTRACT: A type of contract in which costs sustained by the contractor are reimbursed. This type of contract normally is used for R&D tasks—the cost-plus-fixed-fee is a variation of this type of contract.

COST SHARING: A formula used in incentive-type contracts in which a contractor derives a share in any amounts that are less than the target cost that the contract specifies. For any contract costs in excess of target, the contractor absorbs part of the excess costs.

CRITICAL DESIGN REVIEW (CDR): Formal review procedure to establish whether software design satisfies requirements.

CRITICAL PATH: The path of a PERT network which requires the longest time to complete.

CYCLE: A period during which the segments of events are accomplished, leading to a concluding objective. An example is the project procurement cycle, which is the time from contract award to acceptance of final contract item.

DECOMPOSITION: Breaking down of a system design into small modules that are logical, simple, and easily understood; a technique used extensively in the design of computer programs.

DEFENSE ACQUISITION REGULATION (DAR): U.S. government procurement rules (supersedes ASPR).

DEFINITION PHASE: The phase in the configuration management cycle during which the requirements previously established are translated in a specification and associated planning documents.

DETECTION SYSTEM: A system designed to sense different signals and respond in specific ways in accordance with the signal characteristics.

DIGITAL COMPUTER: A machine comprising a multitude of on-off switches that are controlled to execute desired calculations.

DIRECT COSTS: Specific expenses incurred on a project. Such expenses are engineering, manufacturing, and materials used on the project, as opposed to administration, depreciation, etc., which are indirect expenses.

DIRECTIVITY EFFECTS: As applied to radar returns, those effects caused by the position of the reflecting surface relative to the line of signal return to the antenna.

DISCIPLINE SUPERVISORS: The two-boss managers in the matrix who are responsible for the performance of different project teams and who report to both the project manager and the functional manager.

DISPLAY SYSTEM: A system on which acquired information is presented in an intelligible form.

DRAFTING ORDER REVISION: A directive issued to the chief draftsman by the project design engineer to reflect a change in the drawings.

DRAFTING WORK ORDER: A directive issued to the chief draftsman to complete a drafting task.

DUMPING: Transferring all or part of the memory content from one computer section to another or into the input-output section.

EARLIEST EVENT TIME: The earliest date that can be anticipated for the completion of the specified work effort necessary to reach an event.

EFFECTIVE FAILURE RATE: The failure rate of a component as modified by its application, conditions of operation, and environment.

EFFECTIVENESS: A measure of the probability that a piece of equipment will be ready and capable of performing its function.

ENGINEERING ORDER (EO): A directive to the engineering department specifying an engineering task to be performed, its schedule, the budget of cost and hours, and other pertinent information to permit implementation of the engineering task.

ENVIRONMENTAL TESTS: A special series of tests whereby the equipment is subjected to different conditions of operation, such as vibration, shock, and cold.

EVALUATION: An analysis, using set criteria, to establish the relative excellence of documents such as a proposal.

EVALUATION FACTORS: As applied to proposals to be evaluated, the factors represent the criteria used to establish the relative excellence of each proposal in each area.

EVENT: A specific point in a PERT network representing the start or completion of an activity. An event does not have any dimension in time or effort.

EXCUSABLE DELAYS: Failure to meet contract delivery dates due to circumstances beyond the control of the contractor—e.g., natural catastrophes, failure of customer to meet contracted obligations, etc.

EXPECTED ELAPSED TIME: The period of time that is predicted for completing an activity, generally used for PERT.

FINAL SOFTWARE DESIGN REVIEW (FSDR): Review of completed software products, including documentation, testing, integration, and associated records, to establish whether requirements have been satisfied.

FIXED-PRICE INCENTIVE (FPI): A type of contract in which the target price is established with a sharing arrangement between the buyer and the seller for any amounts which the contract cost may fall under or over the contract target cost.

FLOWCHART: A diagram of a computer program which graphically indicates the operational sequence and relationship of functions of a computer program.

FLOW DIAGRAMS: A means of graphically displaying the mathematical model of a system or a computer program design showing data or functional flows.

FORTRAN: A programmed computer language expressed in arithmetic formulae which require translation by the computer into usable machine language.

FOUR-STEP SOURCE SELECTION: A four-step procurement procedure leading to a selection of single source with whom negotiations are held, with the possible result being a contract award.

FUNCTIONAL BASELINE: The point in the configuration management cycle which terminates the definition phase and initiates the acquisition phase.

FUNCTIONAL CONFIGURATION AUDIT (FCA): An audit to verify that the performance of computer program configuration items satisfies the requirements established at the functional baseline.

FUNCTIONAL MANAGER: A manager in the matrix organization responsible for the management of resources for different projects.

FURNISHED PROPERTY: Equipment, data, or other material that the customer is contractually obligated to provide to the contractor in a specific time frame.

GANTT CHART: A graphic presentation in which the data are shown as bars and in which actual versus planned accomplishment is readily observed.

GENERAL AND ADMINISTRATIVE RATES (G&A): Rates applied to the direct and indirect overhead costs of a contract to cover all costs that cannot be identified with any specific program of a company.

GOVERNMENT-FURNISHED PROPERTY: Items furnished by the government to be incorporated into the equipment to be delivered by the contractor or used in fulfilling the contractual requirements.

GRID LINES: A complex of lines, generally vertical and horizontal, which divide an area into small subareas.

HIGH-LEVEL PROGRAM: A computer program which is independent of the machine— FORTRAN and COBOL.

HOUR BREAKDOWN: The hours of engineering, manufacturing, and other types of labor identified with specific tasks or types of effort in a project.

HYBRID SPECIFICATION: A specification for a system that identifies the required performance characteristics and imposes requirements which limit the design approaches that can be used.

IMPLEMENTATION PLAN: A description of the resources and plans an organization proposes to use to reach a particular objective, such as fulfilling the terms of a contract.

INFRINGEMENT: The violation of a right of a party by another party—e.g., patent infringement.

IN-HOUSE DEVELOPMENT: A project undertaken by an organization with its own resources.

INTEGRATED LOGISTIC SUPPORT: General support for equipment used in the field. Support includes provision of spares, instruction, manuals, maintenance, etc.

INTEGRATED TESTS: Tests to verify the performance of several interrelated systems when functioning together as a unit complex.

INTEGRATION: Combining of multiple software elements or software and hardware elements to achieve compatible system functioning.

INTERFACE: The element that provides for translation capabilities to permit exchange of different types of signals and functions between different types of systems (functional compatibility.)

ITERATION RATE: The speed at which a digital computer can perform all the required calculations for a result before repeating its cycle.

LATEST ALLOWABLE TIME: The latest date on which a PERT event can occur without delaying the completion of the program.

LATEST COMPLETION DATE: The latest date on which a particular effort can be completed without effecting a delay in the program completion schedule.

LIFE CYCLE: The period of time from the start of utilization of a piece of equipment until it ceases to have utility.

LINE OF BALANCE (LOB): Management tool for reporting the status of a project or subdivision of a project.

LINE ORGANIZATION: A management structure with a hierarchy headed by an individual, followed by descending levels of managerial personnel.

LOAD MAP: Description of format and location of software modules loaded or stored in a computer system.

LOADING: An amount of effort for which an organization is obligated during a specific period of time.

MACHINE LANGUAGE: A language that does not require translation and that can be used directly in the computer.

MAINTAINABILITY: The ability of a piece of equipment to be maintained or repaired. Maintainability can be expressed by the term "mean time to repair."

MAKE-OR-BUY DECISION: Consideration of different factors to determine whether to purchase a part or system or undertake its design and fabrication with available resources.

MANPOWER LOADING DISPLAY: A graphic presentation of personnel requirements by skill category for specific period integrals.

MATERIAL BREAKDOWN: Identification of components, modules, and material with their costs.

MATRIX EXECUTIVE: Leader of a matrix organization, whose primary responsibility is to ensure the proper function of the matrix organization through the efforts of the project and functional managers.

MAXIMUM LIMIT CONDITIONS: Capacity of a component beyond which failure occurs.

MEANTIME BETWEEN FAILURES: The elapsed time that a particular type of component in a system can be expected to function before failure. The elapsed time is obtained from the probability curve derived from the statistical distribution of previously observed or calculated data.

MEMORY UNIT: A subsystem of a digital computer which functions to store a specified amount of data until required.

NEED TO KNOW: The justification of a person or an organization to obtain access to a particular area or type of information, which is often classified.

NEGOTIATIONS: Discussions between parties regarding the requirements of a proposed contract. The discussions are designed to arrive at a meeting of the minds so that the contract can be formally executed.

NETWORK: A graphic presentation in which the tasks or functions which occur sequentially or in parallel are shown as an array of interconnected lines.

OBJECTIVE CHART: Element of the line of balance report which presents the percent effort of a task that is completed as of the report date.

OPEN: A term generally used in conjunction with electrical work, describing a condition such as a break in a line which results in an interruption in the flow of current.

OPERATIONAL BASELINE: The checkpoint in a configuration management cycle which establishes the equipment and its support item as ready for utilization.

OPERATIONAL PHASE: The configuration management period during which the equipment produced in a project is utilized.

OVERHEAD RATES: A percentage added to a direct task cost, such as manufacturing labor, to cover indirect costs incurred in performing the particular task.

PACKAGING: The physical arrangement of elements of a system—housing, ventilation, and other features of the hardware.

PARETO'S LAW OF DISTRIBUTION: A statistical theory which states that in any system the greatest portion of the total cost is represented by a small percentage of the total number of subsystems.

PERFORMANCE SPECIFICATION: A specification which describes what is to be derived from the equipment but does not describe how the equipment is to be designed or fabricated.

PERT: Program evaluation and review technique.

PHASING: Scheduling of different types of effort to promote an efficient flow of work, particularly for interdependent areas of effort.

PHYSICAL CONFIGURATION AUDIT (PCA): Review of code for computer program contract items (CPIs) to make sure that they are consistent with documentation.

PRELIMINARY DESIGN REVIEW (PDR): Review of design approaches and test plans of the CPCIs.

PROGRAM: A means used in a digital computer to direct and control the functions and operations of the computer.

PROGRAM ANALYZER: Software tool that operates on the computer program to collect data related to performance, testing, coding, and other functions.

PROGRAM EDITOR: Software tool which facilitates the implementation of programming changes and additions by providing for automated formatting and standardization.

PROGRAM LIBRARIAN: Software tool for storing and retrieving program module information that is developed on the project.

PROGRAM LISTINGS: Data relating to a software module which includes language statements, identity and location of instructions, and other relevant information.

PROGRAM TESTING: Testing to verify that software meets performance specifications.

PROGRAM WRITING: The production phase of software development, during which the software design of the system is decomposed into modules, coded, verified, module-tested, and documented.

PROGRESS PAYMENTS: A formula established in a contract which provides for payment to the contractor as various phases are completed.

PROPRIETARY PROCUREMENT: As used in the contractual sense, a sole-source procurement.

QUALITY CONTROL: The implementation of the procedures and means by which the quality of an element or system being produced is maintained and screened.

RADAR SHADOW: The shadow cast by an object as a result of illumination by a source of radar transmission. Any other object in the radar shadow will be obliterated in the resultant radar presentation.

REDETERMINATION: A feature of a contract by which the customer reserves the right to review the costs and rates that are claimed by a contractor and to implement an adjustment in contract price as necessary.

RELIABILITY: The ability of a component, system, or piece of equipment to function in accordance with a specified standard for a particular period of time.

REQUEST FOR PROPOSALS (RFP): A solicitation for negotiated acquisition whereby the government can award a contract without further negotiations.

REQUEST FOR QUOTATION (RFQ): Solicitation to negotiate delivery, price, and other factors with suppliers.

REQUEST FOR TECHNICAL PROPOSALS (RTP): Solicitation for technical proposals used only in the first step of two-step source selection.

RISK FACTOR: A percentage added to the estimate of hours or other costs of a project to cover unanticipated expenditures that might statistically be expected.

ROBUSTNESS: Ability of a computer program to recover from erroneous or inconsistent inputs and continue without detrimental results.

S CURVE: A graphic presentation of some accomplishment as a function of time, the characteristics of which are S-shaped.

SCHEDULE PREDICTION REPORT: A report in which the actual and estimated schedules are compared and the trends of progress are indicated to facilitate schedule prognostications.

SHORT: A term used in electrical design relating to a condition between two points which offers low resistance to the current flow.

SIDE ITEMS: The items of a contract, such as manuals, which are to be used to support the main item of procurement.

SIMULATOR: A system which synthesizes the functions and/or appearance of an operational system, thereby duplicating such equipment.

SLACK: A measure of how much excess time is available or anticipated over and above the scheduled time for the completion of an activity or a series of sequential activities.

SLACK, NEGATIVE: A measure of how much slippage exists or is anticipated in relation to the original schedule for an activity or a PERT path.

SLACK PATH: The measure of slack for a particular path of a PERT network.

SOFTWARE: Computer programs, including associated data and other documents.

SOFTWARE MODULE: A subset of the hierarchy of the system software.

SOFTWARE REQUIREMENTS REVIEW (SRR): Review of criteria for the software to make sure that the system requirements will be satisfied.

SOFTWARE TOOLS: Computer programs which function to provide the automated testing, production, and monitoring of computer software.

SPECIAL-PURPOSE COMPUTER: Specially designed computer providing fixed functions for a particular application or problem.

SPOILAGE: Material spoiled in a manufacturing and fabrication effort that must be discarded as not usable.

STANDARD DEVIATION: A statistical measurement derived from a normal distribution curve. One standard deviation is equal to 68 percent of the total area under the distribution curve.

STATEMENT OF WORK (SOW): A document associated with a procurement which defines requirements for contractor efforts which are not included in the specification.

STRESS RATIO: Ratio of the operating load to the rated load of a component.

SUBSTANTIAL COMPLIANCE: A legal term related to contracts which states that even when certain requirements are not achieved, the contractor cannot be held in default if the primary objectives are acceptable. The contractor would generally have to provide consideration to compensate for the deficiencies.

SUBSYSTEM: An integral portion of a large system that can be functionally separated from the rest of the system.

SUBSYSTEM TESTS: Functional tests on selected portions of a system.

SYSTEM DESIGN REVIEW (SDR): Analysis of system design to make sure that software design satisfies requirements.

SYSTEM ENGINEERING: An overall design approach to make sure that the various subsystems of a design will be compatible and will result in the acceptable performance of the complete system.

SYSTEM INTEGRATION REVIEW (SIR): Review to verify that the integrated functional design of the software and hardware satisfies system requirements.

TARGET COST: The cost figure of a contract that represents the cost norm or goal.

TARGET PROFIT: The nominal profit that a contractor would realize if the contract were to be completed at the target cost figure.

TASK ASSIGNMENT ORDER: A directive issued by the project engineer to complete a specified mission within a planned schedule and budget.

TASK NUMBER: An internal number assigned by the contractor to a program. This number is used for cost accounting, purchasing, production, engineering, and all actions taken in conjunction with the contract.

TECHNICAL PROPOSAL: A technical document presented by an organization proposing means of accomplishing specific objectives.

TECHNICAL PROPOSAL REQUIREMENTS (TPR): A description of the type of information to be presented in a technical proposal, the order of presentation, the format, and other requirements.

TERMINATION: An action that brings all effort on a contract to a halt.

TEST PROCEDURE REPORT: A report containing mutually accepted criteria to be used in the acceptance testing of the procured equipment.

TESTABILITY: Use of simple structure, general algorithms, and logical design that enhances the testing of all functions of a computer program.

TIERS OF DESIGN: Classifications of the degree of design detail of a particular equipment or system.

TIME FUNCTION CONDITIONS: Life span of a component under specific conditions; beyond this span, fatigue failure results.

TOP-DOWN STRUCTURED PROGRAM: The sequential design of a computer program whereby the top module is created and debugged first, using dummy modules for lower levels of the program structure, after which the process is repeated for each successively lower module.

TRACEABILITY: Ability to correlate each element of a software with its associated documentation.

TWO-STEP ACQUISITION: A procurement matter in which step one involves the establishment of which technical proposals are acceptable and step two is the solicitation for bid quotations.

VALIDITY: A program that provides for all the functions necessary to satisfy requirements and to provide for effective interfacing with associated software elements.

VALUE ANALYSIS: A contract by which the savings realized as a result of recommendations made by a contractor are shared between the contractor and the customer.

VERIFICATION TEST: A test to determine whether the equipment is capable of performing as required by the specification or test plan.

WALKTHROUGH: A review of software design or of the code presented by the designer or coder to peer personnel to detect errors, omissions, and other flaws and to appraise the quality of the product.

WORD: A unit of coded information expressed as a series of bits and used in a computer.

WORK BREAKDOWN STRUCTURE (WBS): A "tree" which identifies the different tasks of a project.

INDEX

ABOUT THE AUTHOR

Victor G. Hajek is a project management consultant.
After service with the Navy during World War II, he
worked as an application and sales engineer before
joining the Naval Training Equipment Center
(NTEC). At NTEC he served as project manager for
complex simulators used for the training of antisub-
marine warfare (ASW) crews. Subsequently he was
given responsibilities as branch and division head
over engineers and project managers handling the
wide spectrum of weapon system simulators used for
the training of pilots and crews of ASW, support,
fighter, and other types of military aircraft. In addition, Mr. Hajek headed groups charged with the tasks
of providing engineering systems analysis, long-
range technical and budget planning reports, and engineering consulting services for simulator projects
incorporating visual, computer, and other special
subsystems. During his tenure at NTEC, he realized
that advances in technology, revisions in contracting
procedures, changes in administration and management techniques, and innovations in other related
areas mandated that project managers in all areas of
effort update their knowledge in order to carry out
their assigned functions. The revision of this book
represents one step in that direction.

In the international arena, Mr. Hajek served as
NTEC's representative to NATO on ASW simulators
and as liaison project manager for the Harrier simulator, which was under contract with a firm in England.
Mr. Hajek has conducted seminars on project management subjects in England and is highly regarded
for his informed overview of the complex, multidis-
ciplinary art of coordinating the technical and administrative aspects of engineering projects.